HTD-Vol. 136

FUNDAMENTALS OF PHASE CHANGE: BOILING AND CONDENSATION

presented at
AIAA/ASME THERMOPHYSICS AND HEAT TRANSFER CONFERENCE
JUNE 18–20, 1990
SEATTLE, WASHINGTON

sponsored by
THE HEAT TRANSFER DIVISION, ASME

edited by
L. C. WITTE
UNIVERSITY OF HOUSTON
C. T. AVEDISIAN
CORNELL UNIVERSITY

THE AMERICAN SOCIETY OF MECHANICAL ENGINEERS
United Engineering Center • 345 East 47th Street • New York, N.Y. 10017

Statement from By-Laws: The Society shall not be responsible for statements or opinions advanced in papers . . . or printed in its publications (7.1.3)

ISBN No. 0-7918-0481-X

Library of Congress
Catalog Number 84-72445

FOREWORD

This volume contains papers presented at the 1990 AIAA/ASME Thermophysics and Heat Transfer Conference during the Fundamentals of Phase Change: Boiling and Condensation session. A good mixture of boiling and condensation papers is represented. As is usually the case in such a session, the papers reflect the rich diversity of research that is taking place in phase change heat transfer.

There are two nucleate boiling-related papers, one treating the basic mechanics of bubble formation caused by rapid depressurization (Wang and Bankoff), and the other addressing the novel idea that nucleation might be caused by a secondary mechanism of embryos entrained during bubble bursting (Gopalen and Mesler). Two papers focus on the transition region in boiling. Schroeder-Richter and Bartsch apply the ideas of non-equilibrium thermodynamics to the Liedenfrost phenomenon while Feng and Johannsen provide a generalized correlation for wall superheat during transition boiling in tubes.

Sankaran and Witte present results for highly subcooled external flow boiling over cylindrical heaters; their intent is to see if the transition region might be averted altogether. Nucleate and film boiling data are presented in this paper.

Three of the boiling papers are very practical in nature. Akhanda and James have two papers, one attacking the problem of how the plate characteristics, such as thickness, influences the heat transfer during flow boiling, and the second examining the effect of rib orientation on the enhancement of heat transfer during flow boiling in ducts. Kuehl and Goldschmidt examine a basic problem in refrigeration systems, the behavior of a refrigerant, RF-22, flowing through a capillary tube.

The condensation papers cover the areas of film condensation on curved surfaces and condensation of mixtures. Jacobi, Goldschmidt, Bublitz, and Tree extend the classic Nusselt analysis for laminar film condensation to a hemisphere and consider a variety of boundary conditions for the initial film condensate thickness. Experiments involving condensation on the surface of a sphere are reported by Mahajan, Chu and Dickinson as it relates to the process of mass soldering of electronic components (called "condensation soldering"). They report data for fluids with Stefan number greater than one. The experiments of Hoon and Burnside address the effects of condensation of mixtures on vertical flat plates and horizontal tube arrays. The vapor components were steam and nonane (an immiscible system in the liquid phase): both the azeotrope and nonane rich mixtures were studied.

We were very saddened to learn of the untimely death of one of the authors, Prof. Klaus Johannsen, shortly after he had submitted the final manuscript of the paper by Feng and himself. Those of us who knew him will miss his enthusiasm for phase change research and his innovativeness in developing experimental techniques.

We thank the authors for their patience during the long review process, and also the reviewers, without whom a technical session cannot succeed (a list of the reviewers is included).

Larry C. Witte
University of Houston

C. Thomas Avedisian
Cornell University

REVIEWERS

A. J. Baum
Z. H. Ayub
V. P. Carey
A. M. C. Chan
S. Chandra
S. C. Cheng
V. J. Dhir
A. Faghri
M. Flik
L. R. Glicksman
P. J. Girrantano
P. Griffith
J. Hwalek
R. L. Judd
S. Kandlikar
J. C. Leung
P. J. Marto
H. Merte
B. B. Mikic
M. North
J. Orozco
M. J. Pate
D. A. Pruzan
G. P. Peterson
O. Plumb
T. M. Rudy
V. E. Schrock
N. Shamsundar
E. Ungar
R. L. Webb
G. L. Wedekind
J. W. Westwater
J. C. Yang
J. Zollweg

CONTENTS

BUBBLE GROWTH ON A SOLID WALL IN A RAPIDLY-DEPRESSURIZING LIQUID POOL*

Z. Wang and S. G. Bankoff
Chemical Engineering Department
Northwestern University
Evanston, Illinois

ABSTRACT

The growth rates of bubbles on a solid wall in a rapidly depressurizing liquid pool is of interest for many two-phase flow and heat transfer application. The present work covers the very early stage of void formation and rise in liquid level, before liquid and vapor become well-mixed. bubble growth rates on a vertical stainless steel surface were measured after rapid depressurization. A bubble growth equation, based on the Jones and Zuber (1978) solution for variable pressure, but taking into account the contact with the solid surface is developed.

INTRODUCTION

The growth rate of a bubble on a solid wall in a rapidly-depressurizing liquid pool is of interest for many two-phase flow and heat transfer applications. For example, in a postulated loss-of-coolant accident in a pressurized-water reactor, or the rupture of a pressure-relief diaphragm in a runaway chemical reactor, the liquid, if initially subcooled, will depressurize to a saturated condition, and then begin to form vapor. The two-phase pool level, and thus eventually the discharge flow rate, are affected by the nucleation and growth rates of bubbles in the varying pressure field. In this paper we consider only bubble growth rates. A companion paper will report on bubble nucleation rates. For rapid depressurization, with engineering surfaces in clean water, nearly all the initial bubbles are nucleated on the solid walls, rather than in the bulk liquid. In the first few milliseconds after a large vessel or diaphragm rapture, the pressure drops rapidly. Bubbles form on the walls, but have no time to coalesce, or rise due to buoyancy. The degree of non-equilibrium between the vapor in the bubbles and the bulk liquid determines the bubble growth rates. In the very earliest stage, growth is dominated by surface tension, but this effect becomes negligible before the bubble becomes visible to the naked eye. Nevertheless, this is a very slow growth period, resulting in a time delay between nucleation and significant bubble growth. The next stage is momentum - controlled, since little cooling of the liquid at the bubble wall has as yet taken place. Finally, the diffusion of heat through the thermal boundary around the bubble controls the bubble growth. For constant external pressure, it is then found that the bubble radius increases as the square root of time (Bankoff 1966, 1982).

Bubble growth on solid walls under rapid depressurization presents additional complications, since both the heat transfer rate to the bubble and the vapor density change with time. Furthermore, a very thin liquid microlayer generally exists (if the wall is well-wetted) between the bubble and the wall. The evaporation of this microlayer contributes to the bubble growth. The work reported in the paper consists of an experimental and analytical approach to the problem. Bubble growth rates, while attached to a non-heated, vertical stainless-steel or teflon surface, under rapid depressurization are measured by high-speed motion photography. A bubble growth equation, based on the Jones and Zuber (1978) solution for variable pressure, but taking into account the contact with the solid surface, is developed. The results should be useful in matching the early level swell period to the later quasi-equilibrium period, when the vapor and liquid temperatures are nearly equal, and the relative velocity of the vapor and liquid can be estimated by drift-flux correlations. These results may also be applicable to flashing flow in converging nozzles. A liquid particle experiences rapid depressurization as it passes through the converging section. At some point nucleation and bubble growth begin, and temperature non-equilibrium between liquid and vapor persists for a considerable distance downstream.

*Work supported by the National Science Foundation under Grant MEA-8212483

LITERATURE REVIEW

The problem of vapor bubble growth under constant external pressure is nonlinear and cannot be solved exactly, except under simplifying assumptions which may be inherently contradictory. Thus, a similarity solution was discovered by Birkhoff, et al. (1958), and also Scriven (1959), with the assumptions of (1) zero initial bubble radius (2) no liquid inertia or surface tension. Hence the vapor is always at the saturation temperature at the given external pressure. The surface-tension-controlled and liquid-momentum-controlled phases of bubble growth are thus bypassed, only the late diffusion-controlled stage being considered.

Plesset and Zwick (1952, 1954) and Zwick and Plesset (1955) solved the problem of bubble growth in a superheated liquid from an initial slight displacement from equilibrium, all the way to late diffusion-controlled growth, by matched asymptotic expansions. The late-stage expansion parameter was the ratio of the volume of the thermal boundary layer around the bubble to the volume of the bubble. In order to deal with the moving boundary, the radial coordinate in the convective heat equation in the liquid was replaced by a Lagrangian coordinate. Since the liquid velocity at any point is related to the bubble wall velocity by the continuity equation, the solution for the temperature profile through the thermal boundary layer at the bubble wall, and hence for the bubble volume at any instant of time, immediately becomes implicit. Further, in order to allow an analytical solution of the zeroth-order heat flow problem, a time-like variable was introduced, which contained the bubble radius, and further ensured an indirect, parametric solution. For the late-stage growth, the zeroth-order error was bounded by solving the first-order problem, and bounding the resulting integral. To lowest order, the bubble radius increases as the square root of time, and the surface heat flux varies inversely, as the square root of time. The solution is thus of the same form as the solution for a plane slab subjected to a step-change in surface temperature, except for a proportionality factor. This factor (called a sphericity factor, K_s, by Forster and Zuber (1954)) accounts for the stretching and thinning of the thermal boundary layer as the bubble surface area increases. For the Plesset-Zwick (P-Z) solution compared to the plane-slab solution, $K_s=\sqrt{3}$. Since the plane-slab solution for the surface flux, when a time-varying surface temperature is imposed, is well-known, this suggests that a good approximation to the bubble growth problem may be obtained by multiplying the corresponding plane-slab solution by the constant-surface-temperature sphericity factor. This was the approach adopted by Jones and Zuber (1978) for a bubble in a infinite sea of liquid in a decreasing-pressure field, and followed herein for a bubble attached to a solid with a non-zero contact angle, likewise in a decreasing-pressure field. It should be emphasized, however, that there is no rigorous basis for this approach. These authors used a sphericity factor $K_s=\pi/2$, derived by Forster and Zuber (1954) for the late-time growth of a bubble in an isothermal, constant-pressure liquid. However, the F-Z solution avoided the inherent nonlinearity of this problem by modeling the growing bubble as a spherical heat sink, expanding through a stationary liquid at a rate dictated by mass and energy conservation. The stretching and thinning of the thermal boundary layer owing to convection is thus not present, leading to a growth rate about 10% slower than the P-Z solution. In this work we use the P-Z value, which also gives a better fit to our bubble growth data. Birkhoff, et al. (1958) also obtain the value $K_s=\sqrt{3}$ as a limiting case of their similarity solution for fast-growing bubbles.

The Plesset-Zwick solution was extended by Bankoff (1963) to third order, taking into account lower-order corrections resulting from simultaneously expanding the heat and liquid momentum equation. Skinner and Bankoff (1963, 1964) extended the Plesset-Zwick procedure to spherically-symmetric initial temperature distributions in the liquid, and finally, to arbitrary initial temperature distributions.

Mikic, et al. (1970) found a useful similarity solution for the dimensionless bubble radius, R^+, as a function of dimensionless time, t^+, by interpolating between the limiting cases of pure momentum control and pure heat-conduction control of bubble growth at constant external pressure with uniform initial liquid temperature. This can be verified from their equation

$$R^+ = \frac{2}{3}\left[(t^+ + 1)^{1.5} - t^{+1.5} - 1\right] \qquad (1)$$

for small and large values of t^+. The analysis was also extended to bubble growth on a wall at constant contact angle. Unfortunately, there seems to be no simple way of extending their solutions to time-varying external pressures.

Theofanous, et al. (1976) obtained a numerical solution for bubble growth in a time-dependent pressure field, assuming a quadratic temperature distribution across the thermal boundary layer. The effects of liquid viscosity, liquid inertia, surface tension, interfacial non-equilibrium, and varying vapor density could all be included. Inoue and Aoki (1975), Cha and Henry (1981), and Toda and Kitamura (1983) examined the same problem by introducing coordinate transformation which immobilizes the moving boundary. Inoue and Aoki obtained the interfacial temperature in terms of a convolution integral involving the interfacial heat flux, and also presented an asymptotic solution for the bubble radius in a slowly-varying pressure field by an extension of the P-Z solution. Toda and Kitamura, (1983) employed a thin thermal layer approximation, together with a similarity solution. The agreement between their predictions and their experimental data, obtained by a pulsed laser to form a bubble nucleus, was good when the sphericity factor $K_s = \pi/2$ was included.

Jones and Zuber (1978) solved the bubble growth problem in a variable pressure field by an extension of the Forster-Zuber method (1954), using the same value of K_s. For a linear decay with time of the bubble-wall temperature, the solution obtained was

$$R(t^*) = \left(\frac{\rho_{vo}}{\rho_v}\right)^{1/3} R_o\{1+\frac{2K_s}{\sqrt{\pi}}[Ja_T\, t^{*0.5} + \frac{2}{3} Ja_p\, t^{*1.5}]\} \qquad (2)$$

where t^* is a dimensionless time, given by $\alpha t/R_o^2$, and Ja_T and Ja_p are Jakob numbers for initial superheat and for pressure effects.

Zwick (1960) solved the problem of the diffusion-controlled growth of a vapor bubble in a constant-pressure liquid with internal heat sources. This problem is equivalent to the decreasing pressure problem, with the exception that the vapor density remains constant. Zwick assumed a linear increase of liquid temperature with time. In this case the bubble radius grows initially as $t^{1/2}$, but for long times as $t^{3/2}$.

Tsung-Chang and Bankoff (1986) modified the Zwick solution to account for the system depressurization, and the consequent vapor density variation. Since the relative rate of change of the vapor density is much smaller than that of the bubble volume, an approximate solution can be obtained for the bubble radius vs. time in parametric form. Numerical solutions were obtained, which were compared with the predictions of previous investigations. Burelbach and Bankoff (1987) performed a direct numerical integration of the integral equation expressing the energy balance on the bubble in Lagrangian coordinates.

EXPERIMENT

The experimental apparatus (Figure 1) consisted of a vertical glass tube above which is mounted a measurement section. This was a 25.4 mm I.D. by 100 mm long stainless steel tube, on which a sheathed thermocouple and two charge-type pressure transducers were mounted. Above the measurement section was a 3.8 cm I.D. stainless steel cross with a flange welded on each end. A pneumatic cylinder connected to a stainless-steel cutter was mounted horizontally on one flange. Depressurization was realized by nearly-instantaneous rupture of the 0.127 mm thickness, 38 mm diameter aluminum diaphragm, mounted on the flange opposite to the cutter blade. A satisfactory diaphragm burst was almost always achieved, as shown by the rupture of the diaphragm into four nearly-identical quadrants. The top flange was blocked by a blind flange with its inner surface cut into a conical shape of 18° from the horizontal, so that the shock wave was not directly reflected back into the test vessel.

The test section (Fig. 2) was a 25.4 mm I.D. by 250 mm long glass tube, with a 3 mm thick wall and a hemispherical bottom. For the measurement of bubble growth, the glass tube had a 100 mm long flat surface, in order to avoid optical distortion. A silicone oil bath with magnetic stirrer was used to heat the test section.

A Photek-IV camera was used to take pictures of both the growing bubbles and the liquid level, at framing rates of 5000 per second. A 1 kw high-intensity projection lamp was installed at the rear of the test section about 0.2 m away from the tube center. 100-ft. rolls of Kodak Tri-X reversal black-white film were used for the experiment, and later developed using a Kramer Mark I processor in the laboratory.

After the film ran through a preset 40-foot length and was accelerated to the desired speed, an event control signal was generated by the camera, which simultaneously triggered the solenoid valve, a signal generator, and the computer to begin data acquisition. The pulse from the signal generator activated an internal LED in the camera, producing an initialization mark on one side of the film. The camera itself also produced a 1 KHz timing mark on the other side of the film, which enabled the time for each frame of film, and the corresponding

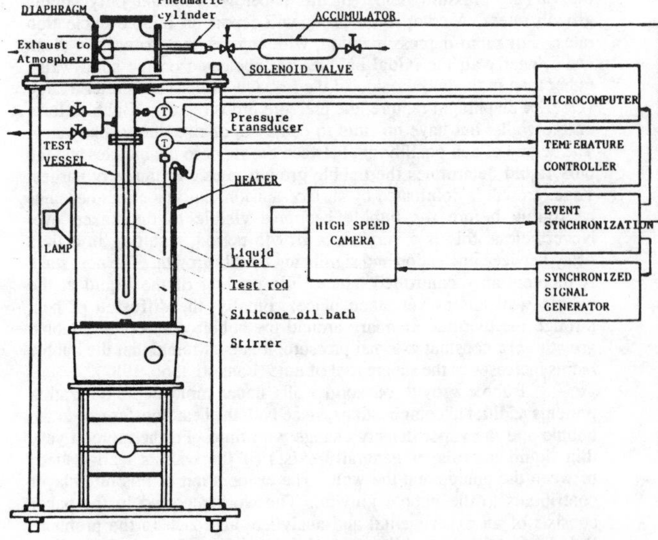

Figure 1. Schematic diagram of the depressurization equipment.

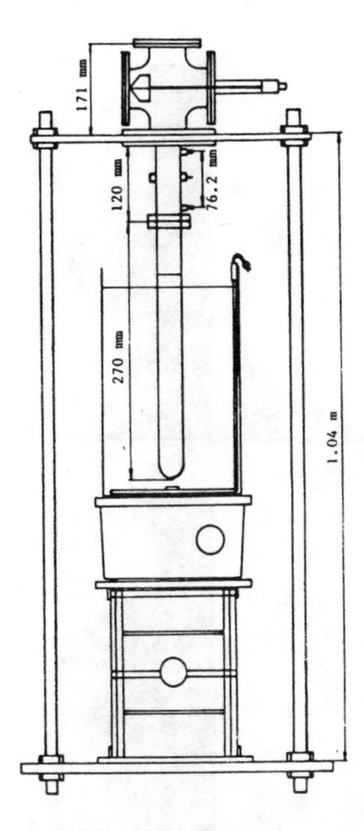

Figure 2. Schematic diagram of the depressurization test section.

pressure, to be determined.

The transient system pressure was measured by two charge-type, high- temperature, high-sensitivity pressure transducers with a response time of 2 μs. To measure the shock wave speed in the g the two transducers were installed 76 mm apart, with water-cooled jackets. The output from the transducers was converted to a voltage signal by a pressure transducer amplifier. The voltage was digitized and stored on a floppy disk. The pressure data file was later transferred to the main-frame computer. The pressure transducers were factory-calibrated, but were recalibrated frequently by discharging nitrogen gas at known pressure from the test section into the atmosphere. Limited adjustments of the pressure transducer amplifier gain coefficient could be made. Factory recalibration was performed when necessary.

A sheathed chromel-alumel thermocouple was used to measure the system temperature. Because the response time of 5 ms of a 0.013-mm diameter unsheathed thermocouple (the smallest commercially available) was of the same order of magnitude as the depressurization time, only the steady state temperature was measured.

The objective was to measure bubble growth rates on a test surface consisting of either a stainless steel or a teflon rod with a milled flat surface. Deionized and distilled water was used as the test liquid. The test vessel was washed with soap several times, soaked in chromic acid cleaning solution for a day, and then flushed with distilled water. Before being installed in the test section, the rod was also washed with soap, put into chromic-acid cleaning solution for an hour, and then washed with distilled water. It was attached to a thin stainless steel bar, clamped between the test vessel and the measurement section. The water in the test vessel was allowed to boil for ten to twenty minutes in order to degas the system. The vent valve was then closed, and the 16mm high speed camera was focused carefully on the center of the test surface. A f=100 mm lens with a 50 mm extension tube was used to take pictures of single bubble growth, and a +4 close-up lens was put on the f/2.8 lens to photograph a column of bubbles. The test rod diameter, or, in some cases, the distance between two artificial holes on the test rod surface, was used as the reference scale.

After the desired temperature (160°C) was reached, the oil bath was removed and the system was further pressurized by admitting nitrogen into the system. After switching on the pressure transducer amplifiers and the projection lights, the camera was started. After developing the film, the bubble diameter and height were measured by projecting the image onto a screen.

ANALYSIS

We consider the problem of a single vapor bubble growing on a vertical wall in an isothermal liquid pool subjected to sudden depressurization. Since the sound velocity in the vapor is large compared to the bubble wall velocity, the pressure within the vapor follows instantaneously its value at the bubble wall, which is given by the equilibrium vapor pressure of the liquid. Furthermore, the liquid can be taken to be incompressible. We assume also that temperature gradients within the bubble can be neglected, although for very fast-growing bubbles this assumption can lead to appreciable errors. For a discussion of these assumptions, see Plesset and Zwick (1954).

Significant complications are introduced by the attachment of the bubble to the solid wall. We assume that the bubble growing on the solid surface is always a truncated sphere, with contact angle, β, and that the center of the sphere is at rest with respect to the bulk liquid. These assumptions, which are made in order to make the problem tractable, are only approximately true. This is because the presence of the solid surface destroys the radial symmetry of the liquid flow around the bubble. In particular, the contact angle depends upon the speed of the contact line, close to the solid wall there is a velocity boundary layer; there is a forward stagnation-point flow around the bubble owing to the translation of the bubble center toward the bulk liquid, and this, in turn, results in slight flattening of the bubble; and buoyancy may cause slight upwards translation of the bubble, although this was not observed in the time scale of these experiments.

In addition, it has been shown that a thin liquid microlayer exists at the base of a vapor bubble growing on a heated solid surface. It seems likely that this is true also for a bubble growing on an unheated wall. We shall assume that the evaporative heat flux to the bubble from the microlayer is uniform over the base of the bubble at any instant and is equal to the heat flux to the curved surface of the bubble at that instant. Further discussion on microlayer heat transfer is given by Tsung-Chang and Bankoff (1989).

With these assumptions, the Rayleigh-Plesset equation in the liquid domain becomes

$$R\frac{d^2R}{dt^2} + \frac{3}{2}\left(\frac{dR}{dt}\right)^2 = \frac{1}{\rho_l}\left(P_v(t) - P(t) - \frac{2\sigma}{R}\right) \tag{4}$$

$$r > R(t),\ 0 \leq \theta \leq \pi - \beta$$

where θ is the polar angle, taking the z-axis normal to the solid wall, and β is the apparent contact angle, assumed to be constant. With these assumptions, the energy balance on the bubble separated from the solid by a very thin liquid layer is

$$\frac{d}{dt}(V\rho_v h_{fg}) = A_b\phi_b + A_c\phi_c \tag{5}$$

ϕ_b and ϕ_c are the heat fluxes through the base and curved surface of the bubble, respectively. The bubble volume is

$$V(t) = \frac{\pi}{3}R^3(t)(2+\cos\beta(2+\sin^2\beta)), \tag{6}$$

and the base area and curved-surface area are, respectively

$$A_b = \pi\,(R(t)\sin\beta)^2, \tag{7}$$

$$A_c = 2\pi R^2(t)(1+\cos\beta),$$

With the above assumptions

$$\phi_c = K_s\phi_b. \tag{8}$$

With these modifications, one can employ either the parametric solution for the bubble radius given by Tsung-Chang and Bankoff (1986), or the plane-slab solution of Jones and Zuber (1978), modified by the constant-pressure sphericity factor, K_s. The former solution has the advantage of not requiring arbitrary assumptions about the correction for convective thinning of the thermal boundary layer around the bubble, but is limited at present to a linear change in bubble wall temperature with time. In addition to the real time, at which the vapor density is known, there is also a time-like variable, $\zeta = \int_0^t R^4(\tau)\,d\tau$. The numerical solution thus involves multiple iterative loops, and is impractical for level swell/nucleation studies, to be discussed elsewhere. A modified Jones-Zuber procedure was therefore adopted.

The surface heat flux of a semi-infinite slab initially at a temperature T_o, which is subjected to a varying surface temperature $T(0,t)=T_o - g(t)$, is given by Carslaw and Jaeger (1959):

$$\phi_b = \frac{k}{\sqrt{\pi\alpha}}\left(\frac{g(0)}{\sqrt{t}} + \int_0^t \frac{g'(\tau)}{\sqrt{t-\tau}}\,d\tau\right). \quad (9)$$

Substituting Equations (6) - (9) into Equation (5) and integrating, one obtains

$$R = \left(\frac{\rho_{vo}}{\rho_v}\right)^{1/3}\left[R_o + \frac{\Lambda}{h_{fg}\rho_{vo}}\int_0^t \left(\frac{\rho_{vo}}{\rho_v}\right)^{2/3}\phi_b(\tau)\,d\tau\right] \quad (10)$$

where the constant factor, $\Lambda = \Lambda(\beta, K_s)$, is

$$\Lambda = \frac{2K_s(1+\cos\beta)+\sin^2\beta}{2+\cos\beta(2+\sin^2\beta)}. \quad (11)$$

We use the Plesset-Zwick value, $K_s = \sqrt{3}$, and take the time, t=0, in Eq. (9) to be the time when significant bubble growth begins. Thus g(0) is the initial superheat for bubble growth, and $g'(t)>0$ during the depressurization. The actual vapor densities, as calculated from the pressure measurements, were used in Eq.(10). Even for this relatively small range of pressures, it is not acceptable to ignore this density correction inside the integral (Jones and Zuber, 1978). A ninth-order polynomial was used to fit the temperature-time curve calculated from the pressure-time data.

RESULTS AND DISCUSSION

After considerable preliminary calibration, about five rolls of 100 ft film were successfully taken for bubble growth measurements. Table 1 lists the measurements for a typical bubble, and Figs 3a-3c

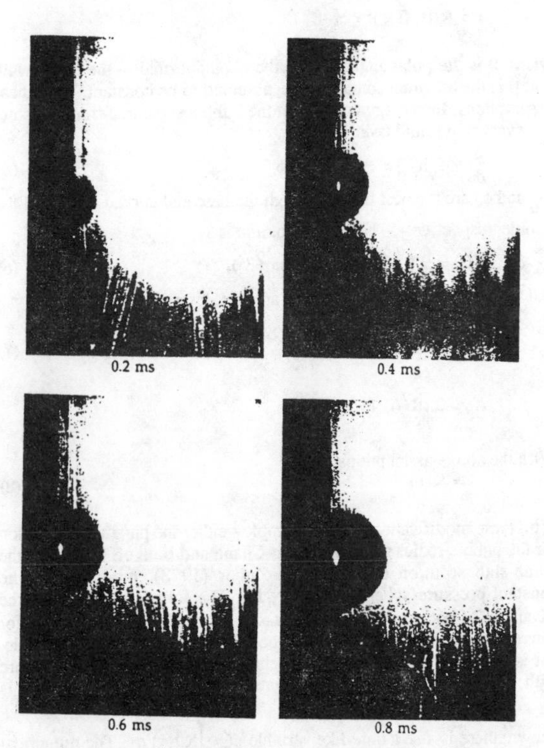

Figure 3a. Bubble shape and size of experiment 3010 at 0.2 ms, 0.4 ms, 0.6 ms and 0.8 ms. In the picture 10 cm equal to 3.76 mm.

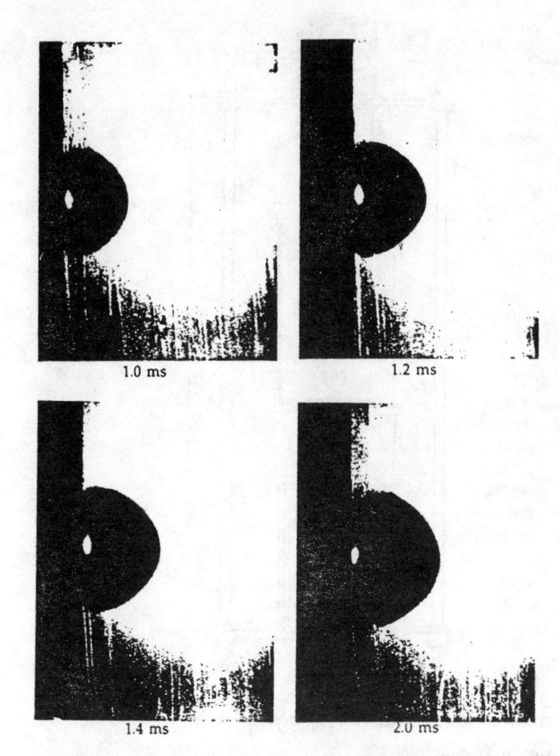

Figure 3b. Bubble shape and size of experiment 3010 at 1.0 ms, 1.2 ms, 1.4 ms and 2.0 ms. In the picture 10 cm equal to 3.76 mm.

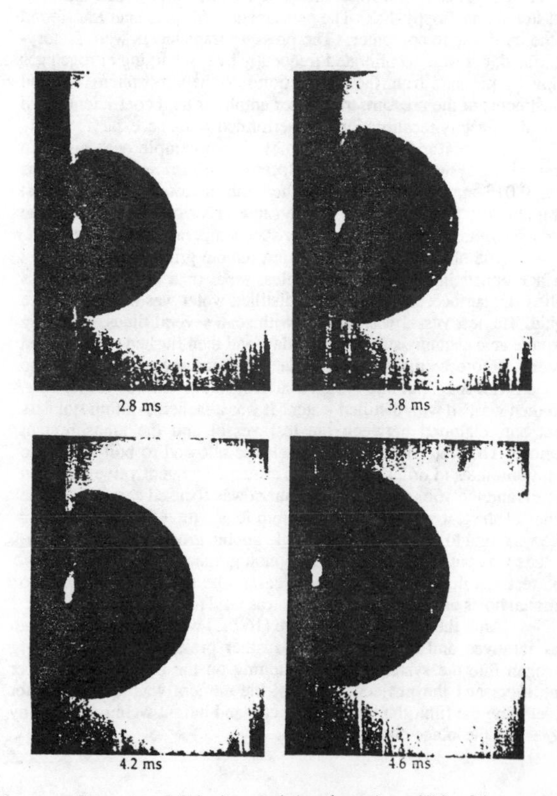

Figure 3c. Bubble shape and size of experiment 3010 at 2.8 ms, 3.8 ms, 4.2 ms and 4.6 ms. In the picture 10 cm equal to 3.76 mm.

show photographs from which the measurements were made. Other bubble measurements are tabulated in Wang (1989). No bubbles were found in the bulk liquid or on the glass walls during the depressurization time period. It can be concluded that for these relatively low liquid temperatures and with the existence of partially-wetted surfaces, heterogeneous nucleation at the test surface, rather than homogeneous nucleation, dominates the boiling process.

The bubble shape, in general, is a truncated sphere with a fairly-constant apparent contact angle, β, in the range 60° ~ 75°. The equivalent-volume spherical bubble radius, R, is tabulated along with the diameter, D, and height, H, of the truncated bubble.

Bubble radii calculated from Eqs. (9) - (11), are plotted in Figs. (4) to (8). As expected, Eq. (1) , due to Mikic, et al.(1970), underpredicts the bubble radius, since it was derived for constant vapor density. Moreover, this equation takes into account inertial, as well as thermal, resistance to bubble growth. As shown by these figures, inertial effects, which slow down bubble growth, do not seem to be significant, since the predicted curves, which are based on purely thermally-controlled bubble growth, are in good agreement with the measured growth curves.

These results can be extended (Wang, 1989) to convert measurements of the rate of rise of the liquid level in the first few milliseconds to rates of bubble nucleation as function of time. Using the pressure-time measurements, one can thus calculate effective nucleus densities and size distributions.

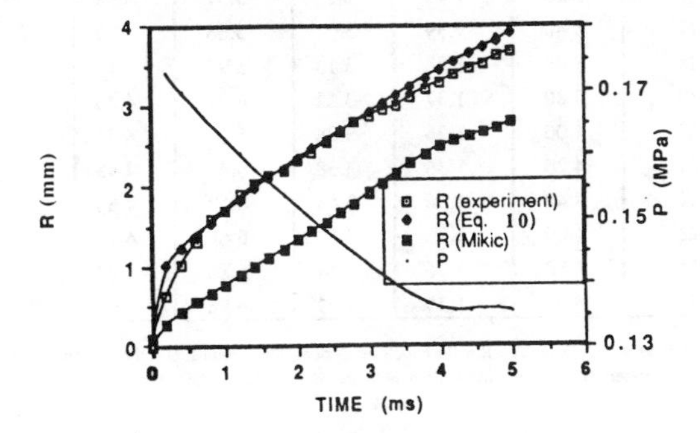

Figure 4. Comparison of experimental bubble growth and theoretical calculation of bubble growth and pressure history during the decompression of experiment 3010.

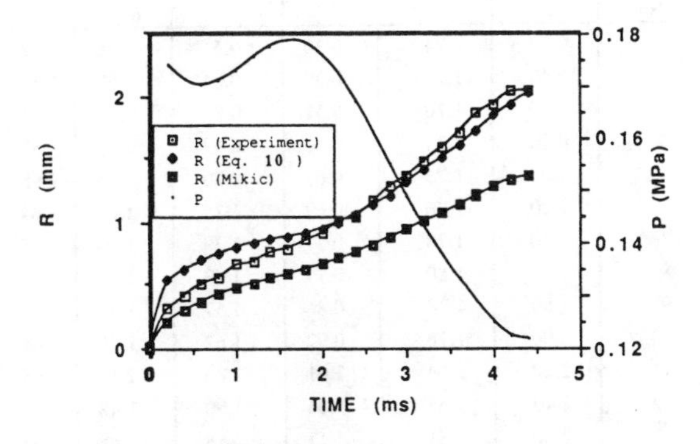

Figure 5. Comparison of experimental bubble growth and theoretical calculation of bubble growth and pressure history during the decompression of experiment 3101.

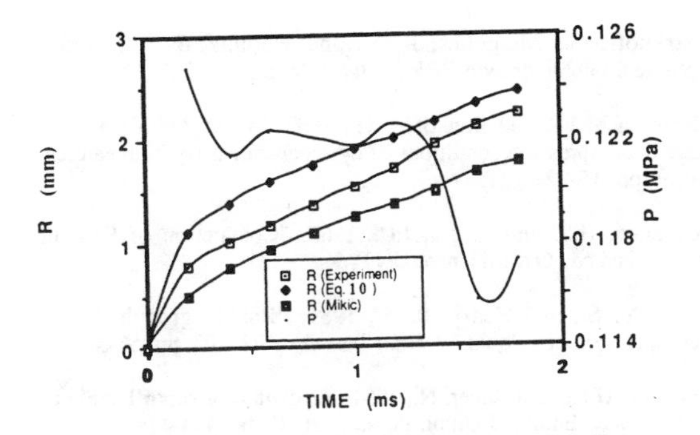

Figure 6. Comparison of experimental bubble growth and theoretical calculation of bubble growth and pressure history during the decompression of experiment 3103.

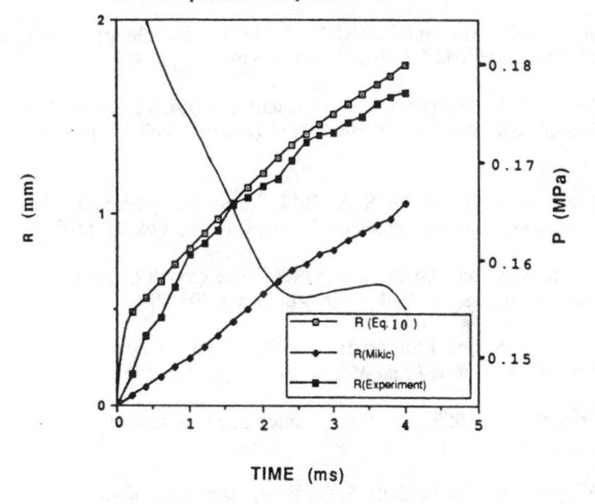

Figure 7. Comparison of experimental bubble growth and theoretical calculation of bubble growth and pressure history during the decompression of experiment 3161.

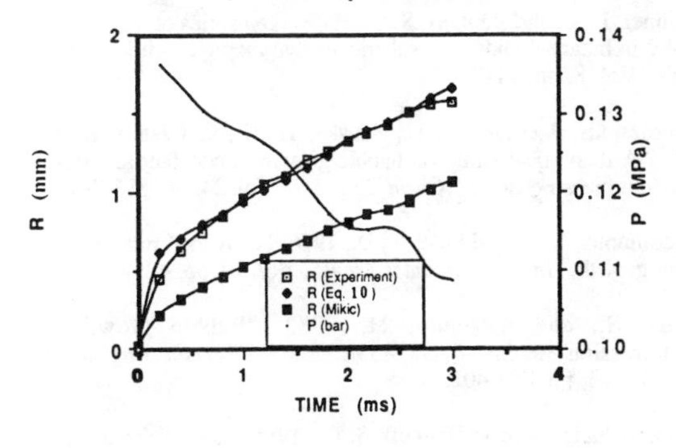

Figure 8. Comparison of experimental bubble growth and theoretical calculation of bubble growth and pressure history during the decompression of experiment 3171.

REFERENCE

Bankoff, S.G., 1963, "Asymptotic growth of a bubble in a liquid with uniform initial superheat," *Appl. Sci. Res.*, Vol. 12, pp. 267-18.

Bankoff, S.G., 1966, "Diffusion-controlled bubble growth," *Adv. in Chem. Eng.*, Vol. 6, pp. 1-60. Ed. by Drew, T.B. and J. W. Hoopes. Jr., Vermeulen, T., and G. R. Cokelet, Academic Press. N.Y.

Bankoff, S.G., 1982, "Lecture Notes on Two-Phase Flow and Heat Transfer," Chap. 8. Los Alamos National Laboratory, Los Alamos, N.M.

Birkhoff, G., Margulies, R. S., and Horning, W. A., 1958, "Spherical bubble growth," *Phys. Fluids*, Vol. 1, pp. 201-204.

Burelbach, J.P. and Bankoff, S.G., 1987, "Vapor bubble growth under decompression conditions", Physicochemical Hydrodynamics, Vol. 9, pp. 15-22.

Carslaw, H.S. and Jaeger, J.C., 1965, "Conduction of Heat in Solids," 2nd ed., Oxford University Press.

Cha, Y. S., and Henry, R. E., 1981, "Bubble growth during decompression of a liquid," *J. Heat Transfer*, Vol. 103, pp. 56-60.

Forster, H.K., and Zuber, N., 1954, "Growth of a vapor bubble in a superheated liquid," J .Appl. Phys., Vol. 25, pp. 474-478.

Inoue, A., and Aoki S., 1975, "On the dynamics of bubble growth under time dependent pressure field," *Bulletin Tokyo Inst. Technology*, Vol. 127, pp. 25-43.

Jones, O.C., Jr., and Zuber, N. J., 1978, "Bubble growth in variable pressure field," *J. Heat Transfer*, Vol. 100, pp. 453-459.

Mikic, B. B., Rohsenow, W. M., and Griffith, P., 1970, "On bubble growth rates," *Int. J. Heat Mass Transfer*, Vol. 13, pp. 657-666.

Plesset, M.S., and Zwick, S. A., 1952, "A nonsteady heat diffusion problem with spherical symmetry," *J. Appl. Phys.*, Vol. 23, pp.95-98.

Plesset, M.S., and Zwick, S. A., 1954, "The growth of bubbles in superheated liquid," *J. Appl. Phys.*, Vol. 5, pp. 493-500.

Plesset, M.S. and Prosperetti, A., 1977, "Bubble dynamics and cavitation," *Ann. Rev. Fluid Mech.*, Vol. 9, pp. 145-185.

Seriven, L.F., 1959, "On the dynamics of phase growth", *Chem. Eng. Sci.*, Vol 10, pp. 1-13.

Skinner, L.A., and Bankoff, S. G., 1964, "Dynamics of vapor bubbles in spherically symmetric temperature field of general variation," *Phys. Fluids*, Vol. 7, pp. 1-6.

Skinner, L.A., and Bankoff, S. G., 1965, "Dynamics of vapor bubbles in binary liquids with spherically symmetric condition," *Phys. Fluids*, Vol. 8, pp. 1417-1420.

Theofanous, T.G., Biasi, L., Fauske, H. K., and Isbin, H. S., 1969, "A theoretical study on bubble growth in constant and time-dependent pressure fields," *Chem. Eng. Sci.*, Vol. 24, pp. 885-897.

Theofanous, T.G., and Patel, P. D., 1976, "Universal relations for bubble growth," *Int. J. Heat mass Transf.*, Vol. 19, pp. 425-429.

Toda, S., and Kitamura, M., 1983, "Bubble growth in decompression fields," *Proc. ASME-JSME Thermal Eng. Joint Conf.*, Vol. 3, pp. 395-402.

Tsung-Chang, G., and Bankoff, S. G., 1986, "Growth of a vapor bubble under flashing conditions," *Proc. 8th Int. Heat Transfer Conference*. San Francisco. Hemishphere Press, N.Y.

Tsung-Chang, G. and Bankoff, S.G., 1989, "On the mechanism of forced-convection subcooled nucleate boiling", *J. Heat Transfer*, to appear.

Wang, Z., 1989, Ph.D. thesis, Chemical Engineering Department, Northwestern University, Evanston, Ill.

Zwick, S. A., 1960, "Growth of vapor bubble in a rapidly heated liquid," *Phys. Fluids*, Vol. 3, pp. 685-692.

Zwick, S.A. and Plesset, M.S., 1955, "On the dynamics of small vapor bubbles in liquids, J. Math and Phys, Vol. 33, pp. 308-330.

TABLE 1

Experimental data for bubble 3010

T_{fo}=403 K

Frame No	t_D ms	P_∞,bar	R, mm	D, mm	H degree	β, deg.
1	0.20	1.72	0.58	1.11	0.74	73
2	0.40	1.70	1.03	1.95	1.28	74
3	0.60	1.67	1.30	2.49	1.67	73
4	0.80	1.65	1.57	3.0	1.99	74
5	1.00	1.63	1.70	3.23	2.24	71
6	1.20	1.60	1.88	3.55	2.49	71
7	1.40	1.58	2.00	3.78	2.67	70
8	1.60	1.56	2.10	3.98	2.77	71
9	1.80	1.54	2.17	4.1	2.88	71
10	2.00	1.52	2.30	4.33	3.06	70
11	2.20	1.50	2.42	4.55	3.23	70
12	2.40	1.48	2.54	4.76	3.41	70
13	2.60	1.46	2.66	5.01	3.54	70
14	2.80	1.45	2.79	5.27	3.7	71
15	3.00	1.43	2.89	5.43	3.91	69
16	3.20	1.41	2.95	5.57	3.93	70
17	3.40	1.39	3.01	5.68	4.02	70
18	3.60	1.38	3.13	5.90	4.18	70
19	3.80	1.37	3.22	6.09	4.26	71
20	4.00	1.36	3.28	6.2	4.36	70
21	4.20	1.36	3.38	6.4	4.48	71
22	4.40	1.36	3.48	6.57	4.64	70
23	4.60	1.36	3.55	6.68	4.73	70
24	4.80	1.36	3.64	6.84	4.87	70
25	5.00	1.35	3.71	6.98	4.97	70

Note: t=0 when $T_{sat}(P_\infty)=T_{fo}$. R is the bubble equivalent-sphere radius; D is the apparent bubble diameter; H is the height of the bubble above the solid surface; and β is the calculated contact angle.

TABLE 2

Experimental data for bubble 3101

T_{fo}=402 K

Frame No	t_D ms	P_∞,bar	R, mm	D, mm	H degree	β, deg.
1	0.20	1.74	0.30	0.5	0.46	57
2	0.40	1.71	0.39	0.64	0.62	54
3	0.60	1.70	0.51	0.90	0.77	61
4	0.80	1.71	0.56	0.99	0.81	63
5	1.00	1.73	0.67	1.20	0.95	65
6	1.20	1.76	0.69	1.25	1.0	64
7	1.40	1.78	0.77	1.39	1.09	65
8	1.60	1.80	0.79	1.43	1.11	66
9	1.80	1.78	0.86	1.57	1.2	66
10	2.00	1.76	0.92	1.67	1.30	65
11	2.20	1.73	1.01	1.85	1.44	65
12	2.40	1.67	1.04	1.90	1.48	65
13	2.60	1.61	1.17	2.13	1.68	65
14	2.80	1.55	1.26	2.31	1.76	67
15	3.00	1.49	1.35	2.45	1.90	66
16	3.20	1.43	1.45	2.69	1.99	68
17	3.40	1.38	1.55	2.87	2.13	68
18	3.60	1.33	1.65	3.10	2..22	70
19	3.80	1.29	1.79	3.37	2.41	70
20	4.00	1.25	1.94	3.55	2.71	69
21	4.20	1.22	2.00	3.74	2.73	69
22	4.40	1.22	2.01	3.74	2.73	69

SURFACE NUCLEATE BOILING, SECONDARY NUCLEATION, AND THRESHOLD SUPERHEAT

B. Gopalen and V. R. Mesler
Department of Chemical and Petroleum Engineering
University of Kansas
Lawrence, Kansas

ABSTRACT

Surfaces of superheated liquids have been observed to burst into nucleate boiling when disturbed. This research was begun to determine whether there exists a minimum superheat, herein called the threshold superheat, that is necessary to initiate nucleate boiling when the surface is disturbed with a stream of droplets of the same liquid. Water and methanol were studied over a range of temperatures, 63° to 86° C for water and 38° to 47° C for methanol. A threshold superheat of 5.5° C was found for water and 14.4° for methanol. The threshold superheat was found to be independent of the saturation temperature. Nucleate boiling penetrated to depths of 15-20 mm with water but only 2-3 mm with methanol. The facts that drops initiated nucleate boiling and that the boiling spread across the surface are interpreted as evidence in support of the secondary nucleation hypothesis.

INTRODUCTION

Evaporation is an important element of chemical engineering. The process is involved in many unit operations in industry, as well as in the laboratory. As its importance warrants, a vast amount of literature has been accumulated about evaporative boiling. In spite of the amount of research that has been devoted to the subject, it is still in a very empirical form, and will likely remain so for a long time to come.

Boiling of liquids occurs in three modes. Calm surface evaporation occurs when the vapor space above the liquid has a pressure that is only a little less than the vapor pressure of the liquid. More vigorous nucleate boiling occurs when the liquid is more superheated and has a steady source of bubble nuclei. Film boiling occurs when the temperature of the heating surface so greatly exceeds the boiling point of the liquid, that a layer of vapor separates the liquid and the heating surface. .

Nucleate boiling has been recognized as the most efficient mode of boiling due to the high heat transfer with only a small temperature difference. It is due to this advantage that nucleate boiling is almost always preferred in industrial boiling systems. This has encouraged a wide interest among the research groups to study nucleate boiling.

Besides the beneficial nature of nucleate boiling, its other side which is harmful in nature has also prompted many research projects to study nucleate boiling. Explosive boiling has a tremendous potential for extensive damage. It has occurred not only in industrial settings but also in the common chemistry laboratories, where the more innocuous bumping is only a milder form of explosive boiling, Mesler (1988).

Perhaps the most important phenomenon in nucleate boiling is bubble nucleation. Bubbles once formed, have a very short life. They usually grow rapidly, rise to a free surface and burst. Therefore a steady source of bubble nucleation is essential in order to sustain nucleate boiling. Three different mechanisms have been mentioned for bubble nucleation. They are heterogeneous nucleation, homogeneous nucleation and secondary nucleation. Heterogeneous nucleation occurs in the imperfections of the heating surface, where a tiny bit of vapor or gas that is entrapped sends a continuous stream of bubbles into the liquid, Cole (1974). This mechanism has long been held to be the only mechanism for bubble nucleation in industrial boiling. Homogeneous nucleation occurs in the bulk of the liquid at temperatures close to the critical temperature of the liquid, Skripov (1974). This mechanism occurs due to the sudden random formation of voids in the bulk of the liquid because of the high kinetic energy possessed by the molecules. The temperature required for homogeneous nucleation to operate is approximately 90% of the absolute critical temperature. Therefore, its occurrence in industrial boiling is rare, and is largely of academic interest.

More recently a third mechanism for bubble nucleation was proposed that operates at the vapor-liquid interface of a superheated liquid, Mesler (1982). In this mechanism, the remnants of the bubble film form drops after the bubble bursts on the free surface that fall back into the liquid. In the process the drops entrain some vapor into the liquid in the form of tiny bubbles. These new bubbles form the nuclei for additional bubbles to grow. The similarity of this process with that of secondary nucleation in crystallization, where the breakdown of the bigger crystals form nuclei for further crystal growth, has lead to the term secondary nucleation in boiling. It is expected that secondary nucleation can be a major contributor along with heterogeneous nucleation for bubble nucleation in industrial boiling, because it operates at temperatures that are commonly encountered in industry. The description of this mechanism and experiments that support it appear in Mesler (1982), Mesler & Mailen (1977), Carroll & Mesler (1981), Esmailizadeh & Mesler (1985), Hall (1987) and Rodriguez & Mesler (1987).

Evidence of secondary nucleation playing a role in nucleate boiling has been inadvertently observed in flash experiments in the past, but secondary nucleation itself went unrecognized. Grolmes & Fauske (1974) conducted some experiments of flash evaporation with water, Freon-11 and methanol. Their glass test sections had diameters of 2 through 50 mm and lengths 10 to 80 cm. They reported the initial liquid superheat required to initiate and sustain two-phase flashing from a free surface as a function of diameter of the test section. During the experiment they photographed the liquid surface with a high-speed movie camera. In their paper they presented pictures where "a single surface bubble grew, shattered, and it appeared that drops from the shattered film caused bubbles to grow as they hit the liquid surface". This is evidence to support the secondary nucleation hypothesis.

Hooper and Luk (1974) conducted flash evaporation experiments in the two limbs of a U-tube, where liquid in one limb was below the saturation temperature at the blowdown pressure while the other was at a higher temperature. When the system was depressurized, flashing occurred only in the superheated limb. They photographed the surfaces in both limbs with a high-speed framing camera to measure the displacement of the liquid levels as a function of time. They observed nucleation just below the surface after depressurization. They proposed a mechanism due to a reaction-type instability, but provided no experimental evidence in support. Their observations too appear to support secondary nucleation.

Friz (1965) studied superheated water in a 7.2 mm dia glass tube in order to understand coolant behavior during reactor accidents. He heated water to temperatures between 105° and 125°C at 3.5 atm, and suddenly reduced the pressure to 1 atm. He noticed that bubbles were formed near the surface, and that the liquid-bubble mixture in the form of an "acceleration front" propagated from the surface into the liquid. There was no clear indication of what initiated boiling once the water became superheated, but clearly the bubble nucleation was not from the solid-liquid interface.

Parrish (1972) conducted flash experiments to observe the pressure transients in the liquid when it is superheated after a rapid depressurization. He presented photographs of the surface after depressurization, taken by a high-speed movie camera, which reveal the initial bubble formation just below the surface before progressing rapidly downwards.

Peterson et al. (1984) conducted flash evaporation experiments. They explain that the absence of nucleation sites within the liquid caused flashing to occur only at the surface. In their experiment they employed a Mach-Zehnder interferometer to study the temperature profiles of the flashing surface. From the interferograms that they present it is evident that the surface and the liquid just below it undergo a rapid cooling, as might be expected in flashing. The interferograms indicate the spreading of bubble nucleation just below the surface over the whole area of the surface. This also appears to be evidence of secondary nucleation.

Secondary nucleation has also been intentionally used to promote nucleate boiling, without recognizing the contribution of secondary nucleation. Hickman (1964) studied the formation of boules, which are drops of liquid that float on the same liquid, but are separated by a shroud of vapor. For this study he required a superheated pool of liquid, which was achieved in a 1 liter Pyrex flask by heating it with a mantle heater. In the construction of his apparatus, he also designed a desuperheater, which was a glass rod with an inverted miniature test tube on the lower end. When this was pushed into the liquid, vapor generated under the open end could only escape as bubbles, initiating ebullition according to the depth and suddenness of immersion. Sudden loss of superheat at initial superheats of 4°-8°C in water was observed by Hickman. His picture depicting the desuperheater in action indicated ebullition that occurred all over the surface. The mechanism for bubbles appearing just below and all over the the surface appears to be secondary nucleation.

Tokmantsev and Chernozubov (1973) conducted an experiment to study adiabatic water flashing. Water was heated in a chamber, and flowed out through a sharp edged aperture in the form of a jet into a flash chamber. They studied the decrease in superheat of the liquid relative to the initial superheat, as a function of the jet length, with the saturation temperature of the flash chamber and the initial superheat as parameters. They also presented pictures of the jet. They observed that individual bubbles appear on the mirror smooth surface at some distance from the aperture before the transparent compact jet breaks into a spray. They also note that fine liquid droplets formed by bursting of the bubbles move not only in the direction of the jet flow, but in other directions too. Clearly in this case there was no opportunity for either the heterogeneous or homogeneous nucleation to operate since there was neither a solid-liquid interface, nor was there a superheat higher than 10°C.

As another part of their experiment, Tokmantsev and Chernozubov (1973) seeded the superheated liquid with fine bubbles by spraying the water surface in the first stage with water drops. They claimed that the falling droplets entrained vapor bubbles, which were incorporated into the flashing jet in the form of the finest bubbles. In this case the flashing began immediately as the jet emerged from the aperture, and the jet lost its superheat at a faster rate than without the spray. This also is evidence of bubble nucleation in support of the secondary nucleation hypothesis.

PURPOSE

After reviewing the various flash experiments in the past, and with the idea of studying the role of secondary nucleation in flash evaporation, it was decided to study the following aspects in flash evaporation.

1. When drops of the same liquid are allowed to fall on a superheated pool of liquid, will they initiate nucleate boiling ? Without the drops, will the superheated pool stay calm without nucleate boiling occurring under similar conditions ?
2. When drops are allowed to fall continuously onto the surface of the pool of liquid as the pool was gradually superheated by lowering the pressure, at what superheat will the drops become effective in triggering the nucleate boiling ? We will call this superheat the Threshold Superheat.
3. Will the nucleate boiling so triggered be confined to a surface layer without reaching the inner bulk of the superheated liquid ?
4. Is this threshold superheat a function of saturation temperature ? Is it a function of the liquid ?

EXPERIMENT

An apparatus was designed such that a quick increase of superheat could be realized in a pool of liquid so that the pool has little opportunity to cool before nucleate boiling could be triggered. This was achieved by depressurizing a flask containing the liquid heated to a uniform initial temperature. The apparatus is depicted in Fig. 1. The liquid was contained in a 3-neck, 1000 ml distillation flask heated in a hot water bath. One of the three necks of the flask was connected to a vacuum reservoir evacuated with an oil-less vacuum pump. A Unimetrics 1160, 0.03 cm OD, stainless steel capillary tube was inserted through the middle neck in such a way that one end of the tube would be about 2.5 cm above the surface of the liquid in the flask. The other end of the capillary tube outside the flask was inserted into a 23 gage needle of a clinical syringe. This arrangement allowed a continuous supply of liquid drops to strike the water surface in the flask.

The third neck was closed by a three-hole stopper. One of the holes held a 23 cm long, 0.25 cm ID glass tube which was sealed on one end. This served as a thermocouple well. Water to a depth of 2.5 cm in this tube improved heat transfer to the thermocouple junction. The second hole held a 5 cm long, 0.25 cm ID glass tube. This glass tube was connected to a pressure transducer by means of a meter length of 0.64 cm polyethylene tube. The third hole held a 6.4 cm long, 0.25 cm ID glass tube that had a plastic burette valve on the outside end. This allowed the flask to be vented to atmospheric pressure after each run.

A T-type thermocouple was used in the thermocouple well to

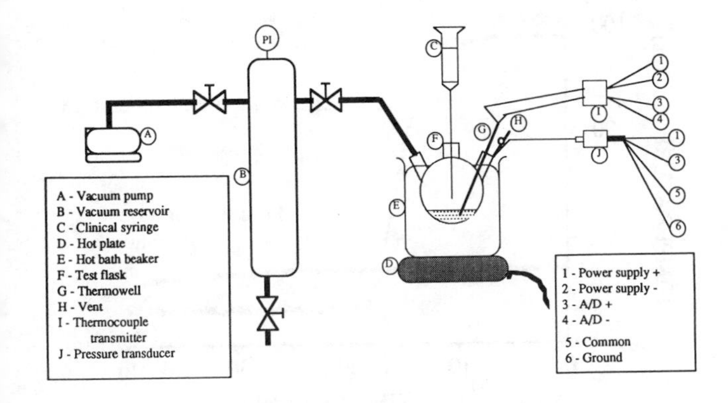

FIGURE 1. EXPERIMENTAL APPARATUS

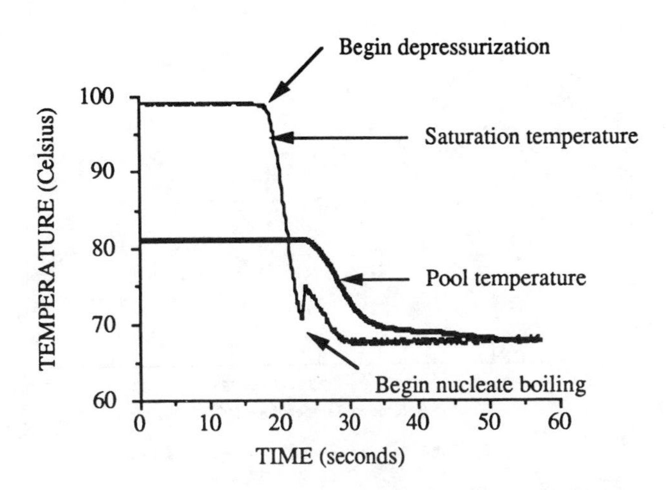

FIGURE 2a : Saturation and pool temperatures for water with addition of drops

measure the temperature of the water in the flask. A Foxboro pressure transducer was used to measure the pressure within the flask. The thermocouple transmitter output and the pressure transducer output were recorded by a PC based data acquisition system.

In order to study the secondary nucleation in the liquid, interference from heterogeneous nucleation had to be eliminated. This was achieved by rinsing the inside of the test flask with nitric acid for approximately 15 minutes before the experiments were started. This treatment was found effective in eliminating nucleating sites. Then, the flask was rinsed with distilled water or methanol thrice to ensure that no nitric acid was left in the flask during the experiment.

Water to be used in the test flask was first boiled at atmospheric pressure for about 30 minutes to degas the liquid and cooled. About 150 ml were poured into the test flask and the flask was then immersed in the hot water bath. The flask was clamped in such a way that it was immersed in the bath and only 1.3 cm of each of the three necks were not immersed.

The three necks were then fitted with their stoppers. The pool temperature was displayed on the computer. The thermostat of the hot plate was set such that the pool temperature reached a desired steady state value. Once the steady state temperature was reached, the vacuum pump was operated until the reservoir pressure corresponded to the desired superheat in the flask. The syringe was then filled with the degassed liquid and fitted to the capillary tube.

A program to take 400 readings of pressure and temperature (600 readings for the first 36 runs with water) at a rate of approximately 11 readings per second was run. The pressure in the flask was quickly reduced by opening the valve on the line connecting the flask to the reservoir. The total depressurization took approximately 5 seconds. Depressurization was gradual, without any sharp changes of the rate. During depressurization, liquid from the syringe was forced through the capillary such that droplets would strike the pool surface at a rate of approximately 3 to 4 drops per second. Water drops had a diameter of 2.3 mm and methanol drops a diameter of 1.8 mm. Once nucleate boiling began, addition of drops was stopped.

Once the 400 or 600 sets of values were taken, the valve to the reservoir was closed, and the vent was opened. The next run was performed after the pool had reached a new steady state temperature.

RESULTS AND DISCUSSION

The pool temperature and the saturation temperature of each run were plotted on the same graph against time. Fig. 2a and 2b show graphs for a typical run with water. The pool temperature curve in Fig. 2a was plotted from the thermocouple readings while the pressure curve in Fig. 2b was plotted from the pressure transducer readings. The saturation temperature curve in Fig. 2a was calculated from the pressure readings

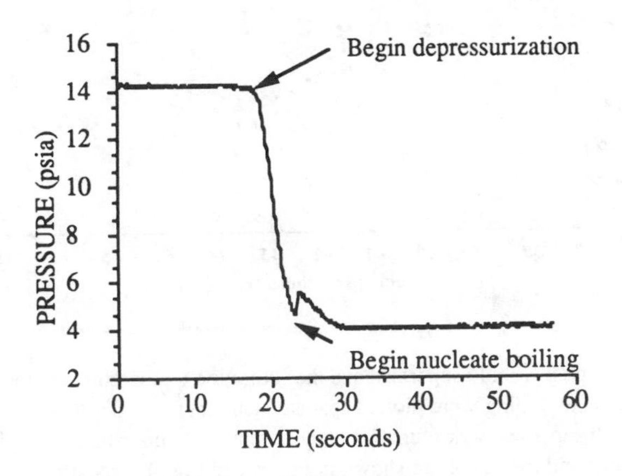

FIGURE 2b : Pressure within flask corresponding to Figure 2a

using Antoine's equation, Smith and Van Ness (1987). Thus the saturation temperature curve in Fig. 2a corresponds to the pressure curve in Fig. 2b.

After depressurization begins the saturation temperature starts decreasing steadily until the boiling mode changes from slow surface evaporation to nucleate boiling. When nucleate boiling starts, the pressure within the flask increases suddenly due to the rapid generation of vapor. The change to nucleate boiling is indicated on the saturation temperature curve by a sudden increase. This increase in the saturation temperature curve corresponds to a similar increase in the pressure curve as seen in Fig. 2b. The difference between the pool and the saturation temperatures corresponds to the superheat present at incipient nucleate boiling. This superheat will be called the threshold superheat. A definite threshold superheat was noticed in all the runs, both with water and methanol.

101 runs were carried out with water in the range 63.5°C to 85.6°C and 100 with methanol in the range 37.6°C to 47°C. Of the 101 runs with water, 87 runs produced acceptable data. The sudden increase in the saturation temperature curve during incipient nucleate boiling was not sharp enough to obtain the exact threshold superheat in the other 14 runs. Hence the data from these runs were not considered for further analysis. In the case of methanol, of the 100 runs, only 97 were considered.

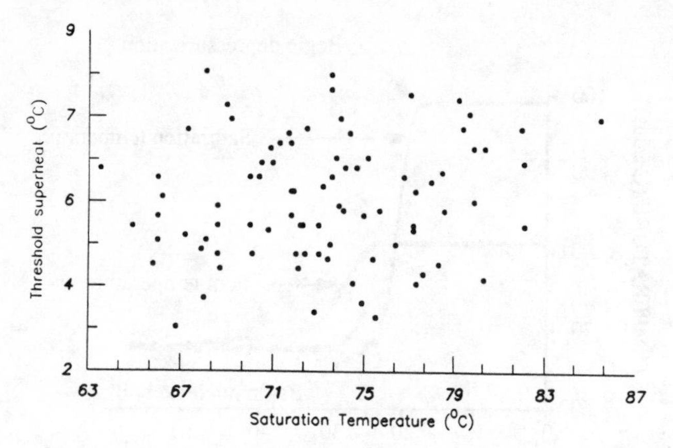

Fig. 3 : Saturation temperature & threshold superheat for water.

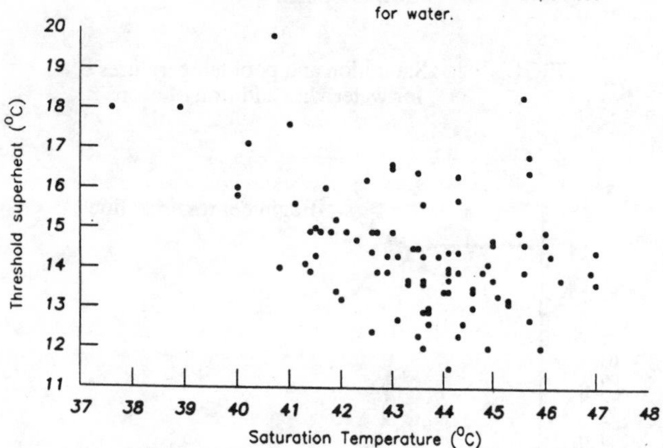

Fig. 4 : Saturation temperature & threshold superheat for methanol.

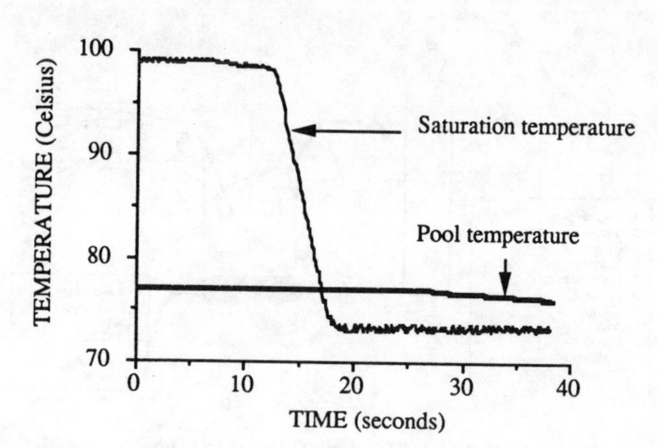

FIGURE 5a : Saturation and pool temperatures for water without addition of drops

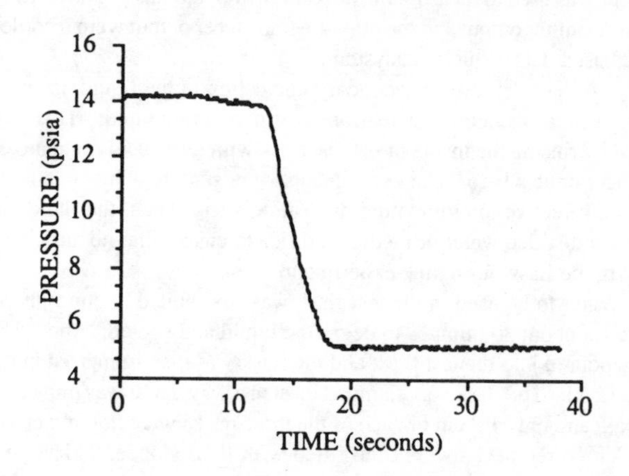

FIGURE 5b : Pressure within flask corresponding to Figure 5a

The threshold superheat and the saturation temperatures at incipient nucleate boiling were plotted against each other in order to see if the threshold superheat was a function of the saturation temperature. The plots for water and methanol are shown in Fig. 3 and Fig. 4 respectively. No functional dependence of superheat on the saturation temperature was observed from the plot by visual observation. The correlation coefficient between the saturation temperature and the threshold superheat was computed for both sets of data. The value of the Pearson's correlation coefficient is 0.17 in the case of water and -0.42 in the case of methanol, confirming the visual observation of the lack of functional dependence of the threshold superheat on the saturation temperature.

A constant value of threshold superheat was observed over the range of saturation temperature, with some amount of scatter. The mean values of the threshold superheat was calculated from the data for water and methanol. For water the mean threshold superheat was 5.5°C + 0.35°C for a 99% confidence level and for methanol it was 14.4°C + 0.40°C.

The mean value of 5.5°C for the threshold superheat for water agrees with Hickman's (1964) observation. He reports that he was able to initiate ebullition in the superheated water with his desuperheater when the superheat was between 4°C and 8°C. Grolmes and Fauske (1974) report the threshold superheat for methanol as a function of the diameter of the test container. When their result was extrapolated to 10 cm dia, comparable to the diameter of the liquid surface in our test flask, the resulting value obtained for the threshold superheat was 13°C, which is in reasonable agreement with the present mean value of 14.4°C.

Runs were carried out in the absence of water/methanol drop addition in order to provide a reference for the study of drops in triggering nucleate boiling. Usually during such runs with water, water droplets condensed on the surface of the flask when depressurization first began due to the adiabatic cooling during expansion. These droplets fell on the surface of the pool, triggering ebullition, whenever the pool was superheated by more than 5°C. Therefore the runs with water had to be conducted with superheats lower than 5°C. Fig. 5a and Fig. 5b are for a run where a superheat of about 4°C that was maintained for about 20 seconds and no ebullition occurred in the absence of drops to trigger nucleate boiling. This is characterized by the absence of the sudden increase in the saturation temperature curve that is normally seen when nucleate boiling occurs, and the absence of a sudden pool temperature drop. The slow surface evaporation is evident from the gradual decrease in the pool temperature after depressurization.

In the case of methanol, nucleate boiling was not triggered because droplets were not formed due to condensation. Therefore much higher superheats of around 20°C were created and maintained for appreciable periods, giving a better indication of the role of drops in triggering nucleate boiling. Fig. 6a and Fig. 6b are for a typical run with methanol without the addition of drops.

In the present experiment drops were allowed to strike the surface continuously as the pressure fell so that the threshold superheat could be estimated accurately. The fact that drops triggered nucleate boiling supports the secondary nucleation hypothesis.

The behavior of the boiling was different for water than for methanol. With water the bubbles were observed deeper in the pool than with methanol. Bubbles were observed at a depth of 15 to 20 mm in water but only 2 to 3 mm in methanol. This behavior also had an effect on the pool

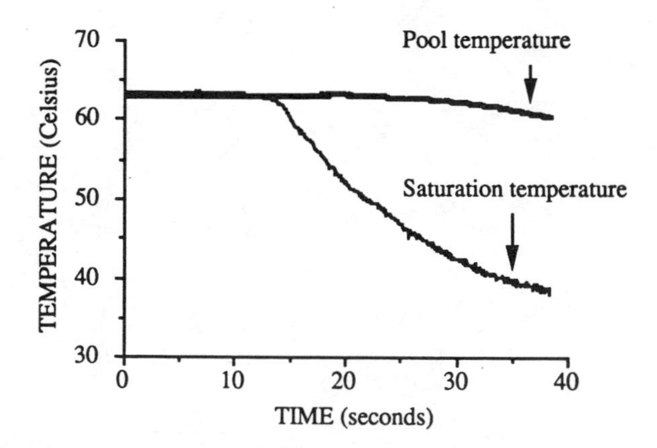

FIGURE 6a : Saturation and pool temperatures for methanol without addition of drops

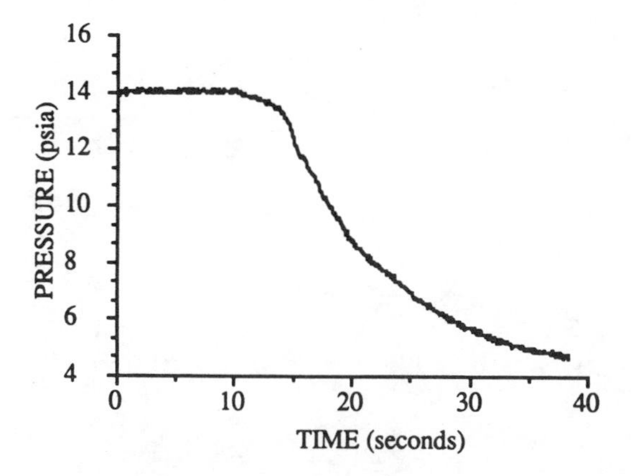

FIGURE 6b : Pressure within flask corresponding to Figure 6a

temperature sensed at a depth of 2.5 cm. With water the temperature always fell to the saturation temperature before nucleate boiling ceased but with methanol the temperature fell only to about 10°C above saturation. These observations indicate secondary nuclei were unable to penetrate more deeply in water than in methanol.

Boiling in thin liquid film is known to be an exceptional case of nucleate boiling. Studies of nucleate boiling from a horizontal surface in which the depth of liquid on the surface was varied have shown that decreasing the depth of liquid improves the heat transfer.

Jakob and Linke (1935) reported evidence of improved performance at low liquid depths when they boiled water on a horizontal surface. Nishikawa et al. (1967) studied distilled water, ethyl alcohol, and an aqueous solution of sodium oleate on several horizontal flat surfaces at liquid depths from 1 to 30 mm. They found that the heat transfer coefficient remains nearly constant provided the liquid depth is maintained above some critical value. Below this critical value, which varies with the liquid, the heat transfer coefficient increases as the depth is decreased. Marto et al. (1977) studied boiling of different liquids on a horizontal surface including distilled water with liquid depths of 25 mm down to thin films near 0.5 mm. They found that the liquid depth had little effect on the nucleate boiling heat transfer coefficient above a depth of 5 mm. Below this depth the heat transfer coefficient could increase by as much as 50% as the depth was reduced. Kim et al. (1983) studied boiling in thin liquid films on a horizontal surface. They report that only at depths less than 7 mm did the depth influence the boiling.

The existence of a critical value for the depth of liquid films, only below which the rate of heat transfer increased as the depth was decreased, could be due to the limited penetration of vapor bubbles of supercritical size as observed in the present experiment. Also the variation of this critical depth for different liquids as observed by Nishikawa et al. (1967) could be the same as the difference in the thickness of the surface layer where nucleate boiling occurred for water and methanol in the present experiment.

A superheat of about 5°C was necessary for water drops added through the capillary tube to initiate nucleate boiling. Once nucleate boiling was initiated however, drops formed from shattered bubbles were able to sustain nucleate boiling at superheats well below 5°C. This possibly indicates that the drops formed from shattered bubbles are more effective in nucleating compared to the drops that were added through the capillary tube.

CONCLUSIONS

When drops of the same liquid strike superheated pools of water and methanol they initiate nucleate boiling if the temperature is above the threshold superheat. Without the drops higher pool superheats can be attained without nucleate boiling occuring.

The mean threshold superheat to initiate nucleate boiling with drops striking the surface was found to be 5.5° C for water and 14.4° C for methanol with some scatter. The threshold superheat does not depend upon the saturation temperature over the range studied.

During nucleate boiling bubbles penetrated more deeply into water than methanol. At a depth of 2.5 cm the temperature fell to the saturation temperature when nucleate boiling ceased with water but with methanol a superheat of 10°C remained when nucleate boil;ng ceased.

ACKNOWLEDGEMENT

This work was supported by NSF Grant CBT-8618879.

REFERENCES

Carroll, K., and Mesler, R., "Bubble Entrainment by Drop-Formed Vortex Rings," *AIChE Journal*, **27**(5), 853 (1981).

Cole, R., "Boiling Nucleation," *Advances in Heat Transfer*, **10**, 85 (1974).

Esmailizadeh, L., and Mesler, R., "Bubble Entrainment with Drops," *J. of Colloid and Interface Science*, **110**(2), 561 (1986).

Friz, G., "Coolant Ejection Studies with Analogy Experiments," *Proc. Conf. on Safety Fuels and Core Design in Large Fast Power Reactors*, **ANL - 7120**, 890 (October 1965).

Grolmes, M.A., and Fauske, H.K., "Axial Propagation of Free Surface Boiling into Superheated Liquids in Vertical Tubes," *5th International Heat Transfer Conference*, **Vol IV**, B1.7, 30 (1974).

Hall, C.M., "Secondary Nucleation in Superheated Water Drops on a Heated Quartz Surface," *Nuclear Energy*, **26**(4), 247 (1987).

Hickman, K.C.D., "Floating Drops and Liquid Boules," *Ind. & Eng. Chem.*, **56**(6), 18 (1964).

Hooper, F.C., and Luk, P.S.K., "The Mechanisms Controlling the Static Pressure in a Flashing Liquid," *5th International Heat Transfer Conference*, **B2.8**, 70 (1974).

Jakob, M., and Linke, W., "Der Warmeubergang beim Verdampfen von Flussigkeiten an Senkrechten und Waagerechten Flachen," *Physik. Zeit.*, **36**, 267 (1935).

Kim, H.-K., Fakeeha, A., and Mesler, R., "Nucleate Boiling in Flowing and Horizontal Liquid Films," *21st National Heat Transfer Conference*, **HTD-Vol. 23**, ASME, 61 (1983).

Marto, P.J., D.K. MacKenzie and A.D. Rivers, "Nucleate Boiling in Thin Liquid Films," *AIChE Sympos. Ser.*, **73**(164), 228 (1977).

Mesler, R.,"Explosive Boiling: AChain Reaction Involving Secondary Nucleation", *ASME Proceedings of the 1988 Heat Transfer Conference*, **HTD-vol 96**, Vol. 2, 487 (1988).

Mesler, R., "Research on Nucleate Boiling," *Chem. Engr. Education*, **XVI**, 152 (1982).

Mesler, R., and Mailen, G., "Nucleate Boiling in Thin Films," *AIChE Journal*, **23**, 954 (1977).

Nishikawa, K., H. Kusuda, K. Yamasaki and K. Tanaka, "Nucleate Boiling at Low Liquid Levels," *Bulletin JSME*, **10**, 328 (1967).

Parrish, J.V., "An Experimental Study of the Initial Pressure Transients During the Flashing of Superheated Water," MS Dissertation, Department of Chemical and Petroleum Engineering, University of Kansas (1972).

Peterson, R.J., Grewal, S.S., and El-Wakil, M.M., "Investigation of Liquid Flashing and Evaporation Due to Sudden Depressurization," *Int. J. Heat Mass Transfer*, **27**(2), 301 (1984).

Rodriguez, F., and Mesler, R., "The Penetration of Drop-Formed Vortex Rings into Pools of Liquid," *J. of Colloid and Interface Science*, **121**(1), 121 (1988).

Skripov, V.P., " Metastable Liquids," John Wiley, New York (1974).

Smith, J. M and Van Ness, H. C., "Introduction To Chemical Engineering Thermodynamics," 4th Ed., McGraw-Hill, New York, 183 (1987). Tokmantsev, N.K., and Chernozubov, V.B., "Investigation into Adiabatic Water Evaporation in the Test Flash Distillation Chamber," *4th International Symposium on Fresh Water from the Sea*, **1**, 497 (1973).

THE LEIDENFROST PHENOMENON CAUSED BY A THERMO-MECHANICAL EFFECT OF TRANSITION BOILING: A REVISITED PROBLEM OF NON-EQUILIBRIUM THERMODYNAMICS

D. Schroeder-Richter and G. Bartsch
Department of Chemical and Energy Engineering
Technical University of Berlin
Berlin, Federal Republic of Germany

ABSTRACT

During transition boiling the solid is partially wetted while the temperature of the wetting liquid is considerably higher than the temperature of saturation at a corresponding pressure in the fluid center. Since recently constant pressure within the fluid has been assumed, only the additional assumption of meta-stability by Spiegler could explain a liquid state. Our fundamentally new approach is to assume both, liquid and vapor near the solid in saturated states at differrent pressures. The Leidenfrost phenomenon seems to occur at approximately adiabatic condition i.e. at low heat flux. Therefore the enthalpy of evaporation will be provided by the mechanical energy of the depressurized liquid alone.

NOMENCLATURE

h = specific enthalpy, J/kg;

$\bar{h}$ = mean value of the specific enthalpy in the cross section of a given axial position, J/kg;

h_g, h_w = specific enthalpy of the vapor and the wetting liquid at the heated surface, J/kg;

$h_{fg} = h'' - h'$ = specific enthalpy of evaporation at constant pressure, J/kg;

m = evaporation mass flux through the interface per unit area and unit time, kg/(m²s);

M = molar mass, kg/kmol;

p_g, p_l, p_w = objective thermodynamic pressure in the vapor, in the center of the subcooled liquid and in the wetting liquid at the heated surface, Pa;

p_s, p_{Leid} = saturation pressure in general and especially corresponding to the Leidenfrost temperature, Pa;

q_g, q_w = heat flux per unit area of the vapor-liquid interface related to the interface system of coordinates, input into the vapor and output from the liquid, W/m²;

q^*_g, q^*_w = heat flux per unit area of the vapor-liquid interface transformed into the vapor system of coordinates, input into the vapor and output from the liquid, W/m²;

$R = 8314$ J/kmol K = gas constant;

$S = w_{ag}/w_{al}$ = slip ratio (dimensionless);

T_g, T_l, T_w = temperature in the vapor, in the subcooled liquid center and in the wetting liquid at the heated surface, K;

T_s, T_{Leid} = saturation temperature and Leidenfrost temperature, K;

u_g, u_w = specific internal energy of the vapor and the wetting liquid at the heated surface, J/kg;

v = specific volume, m³/kg;

w_g, w_w, w_σ = velocity normal to the phase interface for the vapor, for the liquid and for the interface, m/s;

w_{ag}, w_{al} = axial velocity of vapor and liquid, m/s;

X, X_{eq} = true quality and thermodynamic equilibrium quality (dimensionless).

Greek Symbols

α_{eq} = thermodynamic equilibrium void fraction (dimensionless);

ρ_g, ρ_w = density of the vapor and the wetting liquid at the heated surface, kg/m³.

Subscripts, Superscripts and Abbreviations

The superscripts "and' designate the satured vapor and liquid state.

DNB Departure from Nucleate Boiling ;
OBD Onset of Bubble Detachment ;
ONB Onset of Nucleate Boiling.

INTRODUCTION

If the temperature of a heated solid surface, wetted by a liquid fluid, increases somewhat about the temperature of saturation, some vapor bubbles will be produced, called "Onset of Nucleate Boiling" (ONB). With a further increase of the solid temperature the rate of bubble production increases as well, and thereby the transfered heat from the solid to the two-phase fluid increases by the square of the superheat of the solid. This is governed by the growth of the absolute enthalpy necessary for phase transfer. This relationship was correlated empirically by Thom et al. (1965) and deduced analytically by Bartsch & Schroeder-Richter (1989). Schroeder-Richter & Bartsch (1990a) have shown that the maximum heat transfer during forced upflow nucleate boiling and pool boiling is reached, if the relative velocity of phase transfer equals the speed of sound, either for the liquid (velocity on top of the bubble related to the liquid microlayer on the base) or for the vapor phase (velocity of the vapor inside the bubble related to the liquid anywhere outside the bubble). This situation is called "Departure from Nucleate Boiling" (DNB). From now on the transfered heat decreases with a further increase of the solid temperature which is in turn governed by a decrease of the mean area that rests in wetted state related to the total surface of the solid. This situation is called "Transition Boiling" since a process governed by the heat flux will be unstable and either the heated surface will be cooled back to nucleate boiling or heated up to stable film boiling. In the latter case the boiling process will regain stability if the solid is completely unwetted. That occurs at a special temperature of the solid named after the German medical Doctor Leidenfrost (1756), who first described this phenomenon.

A Summary of Recent Investigations

The solid is partially wetted during transition boiling even though the wall temperature, and thereby the temperature of the wetting liquid, is considerably higher than the temperature of saturation corresponding to the pressure in the fluid center. Since recently constant pressure within the fluid has been assumed, a liquid state can only be explained with the additional assumption of a meta-stable region within the diagram of state.

Developing a thermodynamic model describing the Leidenfrost phenomenon Spiegler et al.(1963) pointed out that is was possible to apply the model of Meyer (1911) which explaines the region of meta-stable liquids within the diagram of state (pressure versus specific volume) using the Van der Waals (1873) equation of state. An isothermal function, pressure depending on specific volume, shows two extreme values within the diagram of state. The fitted lines between minima and maxima of different isothermal functions are called spinodal lines. Meyer (1911) obtained the spinodal of the minima to be the limiting line of meta-stable liquids and Spiegler et al. (1963) adopted these lines as graphical description of the Leidenfrost temperature within the diagram of state (foam limit).

The existence of meta-stable liquids described by the spinodal seemed to be proved by Lienhard & Schrock (1966), Skripov (1972), Blander & Katz (1975) and many other empirical investigations. However it was pointed out that the Van der Waals (1873) equation does not fit all kinds of fluids to the same degree of accuracy. Modifications (Carbajo, 1985) have been proposed to take into account different equations of state (i.e. Juza, 1965, Himpan, 1955 and Biney et al. 1986) as well as the properties of the heated surface to calculate the Leidenfrost temperature (Baumeister & Simon, 1973). Bjornard & Griffith (1977) added the latter effect to the so-called "homogenous Leidenfrost temperature" obtained by Spiegler et al. (1963) to estimate the "heterogeneous Leidenfrost temperature". The difference in the mean temperature at the heated solid (heterogenous temperature) and the mean temperature of the wetting liquid (homogenous temperature) is governed by the fluctuations of the wetting zones of these two semi-infinite slabs and has been calculated using an expression of Carlslaw & Jaeger (1959). Additionally Gunnerson & Cronenberg (1978) used special equations of state to describe the deviating behavior of liquid metals.

Further approaches by Lienhard (1976) and Lienhard & Karimi (1978) to correlate the so-called spinodal limit directly using a corresponding-states method do not make use of any assumption dealing with meta-stable states. Nevertheless this correlation fails at moderate and intermediate pressure (Nishio, 1987) and it fails as well when calculating the Leidenfrost temperature for liquid metals (Gunnerson & Cronenberg, 1978). Thus Nishio (1987) correlates the Leidenfrost temperature separately for the three pressure regions low, intermediate and high. On the other hand it seems to be doubtful that the Leidenfrost phenomenon should be governed by three different effects. Therefore the existence of meta-stable states could be questionable.

Competing with the thermodynamic models, some hydrodynamic theories (Berenson, 1961, Gunnerson & Cronenberg, 1980) have been developed based on the stability analysis of Zuber (1958). A suitable heat transfer correlation is used to obtain the minimum film boiling temperature from the (theoretically) well known minimum heat flux. Unfortunately this analytical model predicting the minimum heat flux has not been veryfied sufficiently since steep temperature gradients in the solid occuring together with the Leidenfrost phenomenon hamper the necessary accuracy of the measurements. Using the heat transfer correlation this approach for calculating the Leidenfrost temperature is empirical. In addition the hydrodynamic models do not explain why the superheated solid is partially wetted.

On the other hand thermodynamic models could be explained since van Hove (1950) has deduced from statistical thermodynamics that meta-stable thermodynamic states are indeed able to survive in the case of a one-dimensional chain of atoms or molecules; but in the multi-dimensional case van Hove (1949) has shown that small fluctuations of thermodynamic properties lead to spontanous phase transition. Thus meta-stable states cannot survive the perturbations of the boiling process. Moreover, in presenting the laws of non-equilibrium thermodynamics Prigogine & Defay (1950) noted that meta-stable thermodynamic states can only survive if different conditions of stability hold.

Comparing these conditions with the occurance of turbulences during boiling, Schroeder-Richter & Bartsch (1990b) assumed that none of these stability conditions requested by Progogine & Defay (1950) hold. In spite of this Schroeder-Richter & Bartsch (1990b) have shown that the empirical investigations of Lienhard et al. (1978) and Alamgir et al. (1980) can also be explained by a mechanical non-equilibrium between the vapor and liquid phases. The evaporation of bubbles seems to be governed by a thermo-mechanical effect.

MODEL AND ANALYSIS

During the phase transition the local density of the fluid element varies considerably. Therefore a momentum results at the interface which depends on the velocity of the vapor produced with respect to the liquid left behind. Consequently Schroeder-Richter & Bartsch (1990c) assumed that the wetting liquid "w" near the solid is not superheated and thereby not meta-stable but saturated at a higher pressure p_w than the saturated vapor "g"; *) the pressures in the vapor p_g and the center of the fluid p_l are assumed to be equal:

$$p_w > p_g; \tag{1}$$

$$p_g = p_l \;. \tag{2}$$

If the saturation temperatures of the vapor

$$T_g = T_s\,(p_g) \tag{3}$$

and the wetting liquid near the wall

$$T_w = T_s\,(p_w) \tag{4}$$

are known, we can use the ideal gas approach of the Clausius-Clapeyron equation to calculate a considerable mechanical non-equilibrium

$$p_w = p_l \cdot \exp(h_{fg}\, M/R \cdot (1/T_s(p_l) - 1/T_w)), \tag{5}$$

where h_{fg}, M and R designate the enthalpy of evaporation, molar mass and ideal gas constant.

However this model has recently been successfully applied to calculate nucleate boiling up to DNB (Schroeder-Richter & Bartsch 1990a). In the present investigation we postulate that the thermo-mechanical effect holds during transition boiling and when finally discussing the results obtained we will see that it is useful to postulate this.

First let us consider vertical forced-convection boiling. Recent investigations of Feng & Johannsen (1989) and Weber & Johannsen (1989) dealt with temperature governed steady state transition boiling of sub-cooled water with forced upflow through a small copper cylinder. Interpreting the flow regimes and thereby the location where the Leidenfrost phenomenon is assumed to occur we obtain the following flow model (cf fig. 1).

At the location where the Leidenfrost phenomenon is assumed to occur we postulate adherent vapor at the heated solid surface. According to the previous flow model for the nucleate boiling region (Schroeder-Richter & Bartsch 1990c) we assume that the vapor is at the temperature of saturation, i.e. eq. (3). Indeed this implies a temperature jump between the heated solid surface, which is assumed to be at the Leidenfrost temperature and the adherent vapor film. On the other hand the Leidenfrost phenomenon defines a singular point only where a stable vapor

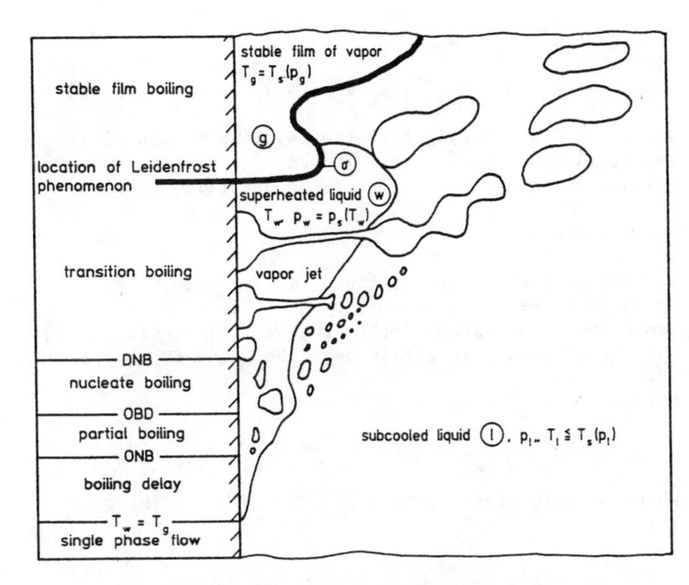

Fig. 1 Model of flow regimes of separate phase flow with time dependent interfaces (upflow)

film is initialized. Now the vapor could be heated up to the temperature of the solid by means of the poor heat conduction of the vapor phase alone. Superheated vapor will thus be expected at any downstream position.

Nevertheless with the production of a stable vapor film the pressure on the dry solid surface is assumed to be equal to that of the subcooled center of the flow $T_l < T_s(p_l)$, according to eq. (2). The wetting liquid in the transition boiling region at any upstream position is assumed to be saturated at a higher pressure, according to eqs. (4) and (5).

At the location where the Leidenfrost phenomenon occurs an interface σ between the vapor phase g and the wetting liquid phase w is expected. We obtain the following equations of balance as jump conditions at the interface σ (Müller, 1985):

mass:

$$\rho_w(w_w - w_\sigma) = \rho_g\,(w_g - w_\sigma)\,, \tag{6}$$

momentum:

$$p_w + \rho_w(w_w - w_\sigma)\,w_w = p_g + \rho_g(w_g - w_\sigma)w_g \tag{7}$$

and energy:

$$(u_w + w_w^2/2)\,\rho_w(w_w - w_\sigma) + p_w\,w_w + q_w = \tag{8}$$

$$(u_g + w_g^2/2)\,\rho_g(w_g - w_\sigma) + p_g\,w_g + q_g,$$

where ρ,w and u designate the density, the velocity, (normal to the interface) and the specific internal energy. The heat flux q depends on some further definitions (Muschik & Müller, 1983). Usually the balance of energy (8) is written in order to describe membrane processes (Førland et al. 1988). Therefore the membrane itself describes an objectively fixed point for the system of coordinates. However in a colloid system, in which a boiling process could be understood, the whole interface of each bubble is moved in all directions of space. Thus the bari-center of the bubble seems to describe a point where a new system of coordinates could be objectively fixed in the same manner as Prigogine (1947).

Consequently we can introduce the corresponding heat flux by a transformation of coordinates (Müller,1985)

*) Nevertheless we use the classical expression "superheated" in order to make the article easier to read.

$$q_g^* = q_g + w_g\,(w_g - w_\sigma)\,.\,\rho_g(w_g - w_\sigma) \quad (9)$$

$$q_w^* = q_w + w_w\,(w_g - w_\sigma)\,.\,\rho_w(w_w - w_\sigma), \quad *) \quad (10)$$

where $(w_g - w_\sigma)$ designates the relative motion of the two systems of coordinates. In order to rewrite the balance of energy we introduce the specific enthalpy

$$h = u + p/\rho \quad (11)$$

and using equations (8) through (10) we obtain the enthalpy of evaporation corresponding to different temperatures of saturation in the wetting liquid and the adherent vapor film near the Leidenfrost point:

$$h''(T_g) - h'(T_w) = (w_w^2 - w_g^2)/2 + w_\sigma(p_w - p_g)/m + (q_w^* - q_g^*)/m + (w_g - w_w)\,(w_g - w_\sigma), \quad (12)$$

where m designates the mass flux through the interface:

$$m = \rho\,(w - w_\sigma) \quad (13)$$

according to eq. (6). The super-scripts "and' refer to the saturated vapor and liquid state. Using the balances of mass (6) and momentum (7) we get:

$$(p_w - p_g)/m = (w_g - w_w) \quad (14)$$

and thus eq. (12) reads:

$$h''(T_g) - h'(T_w) = -\,(w_w + w_g)\,.\,(p_w - p_g)/2m + w_\sigma\,(p_w - p_g)/m + (q_w^* - q_g^*)/m + (w_g - w_\sigma)(p_w - p_g)/m = (w_g - w_w)\,(p_w - p_g)/2m + (q_w^* - q_g^*)/m. \quad (15)$$

Let as now consider the jump in the heat conduction $q_w^* - q_g^*$ which is required to transform the liquid into the vapor phase. Thus the vapor is usually assumed to have a higher level of energy. During transition boiling there is a delay during which evaporation of the liquid near the wetted solid occurs. This is due to the heat flux which is required to submit the enthalpy of evaporation. In constrast, at the location where the Leidenfrost phenomenon occurs we assume that the evaporation occurs promptly. That means that necessary enthalpy of evaporation is available within the mechanical energy alone and heat conduction approaches zero:

*) When obtaining the system of coordinates for the special boiling process (i.e. the bary-center of the vapor) we define the objective kinetic energy for both systems, vapor and liquid. The present investigation deals with unmoved vapor (i.e. onset of stable inverted annular flow). Thus the liquid, which is pushed away from the interface (i.e. the source of the specific volume of vapor), carries the whole kinetic energy.

In the case of spray cooling the impact velocity of the droplets will contribute for the enthalpy of evaporation as well, but the evaporation needs additional kinetic energy to blow the produced vapor away from the interface (i.e. the source of specific volume). Thus eqs. (9) and (10) cannot be applicable for spray cooling.

Nevertheless Testa & Nicotra (1986) have observed bubbles within drops which were partially wetting a heated molybdenum surface. Thus eqs. (9) and (10) should hold for the adherent bubbles within the drops as well (i.e. the true Leidenfrost experiment). This

$$q_w^* - q_g^* = 0. \quad +) \quad (16)$$

At this adiabatic condition and using the definition (13), the balance of energy (15) reduces to:

$$h''(T_g) - h'(T_w) = 1/2\,.\,(1/\rho_g - 1/\rho_w)\,.\,(p_w - p_g). \quad (17)$$

Eq. (17) is our new definition of the Leidenfrost point. In order to get a more useful expression we introduce the specific volume

$$v = 1/\rho \quad (18)$$

and rewrite the thermodynamic properties at the two saturated states (3) and (4):

$$h''(T_g) - h'(T_{Leid}) = 1/2\,.\,(v''(T_g) - v'(T_{Leid}))\,.\,(p_s(T_{Leid}) - p_s(T_g)). \quad (19)$$

Now we have written an equation comparing the properties of two saturated states which depend on the two temperatures T_{Leid} and T_g alone. Thus, if the temperature T_g is given by the system pressure p_1 with $T_g = T_s(p_1)$, equation (19) will hold for a special temperature $T_w = T_{Leid}$ only, which can be calculated iteratively using a table of state:

$$T_w = T_{Leid} = f(T_g) = f(p_1). \quad (20)$$

We obtain this temperature instead of the so-called "homogenous Leidenfrost temperature" calculated by the model of Spiegler et al. (1963). We call it a thermo-mechanical Leidenfrost temperature, since necessary enthalpy of evaporation is available within the mechanical energy alone, if a thermo-mechanical effect, eq. (5), can be postulated. Additional corrections to obtain the so-called "heterogenous Leidenfrost temperature" can be applied analogous to Bjornard & Griffith (1977) or Carbajo (1985) but are not considered here.

However, in order to calculate the thermo-mechanical Leidenfrost temperature using equation (19) the following considerations could simplify the iteration procedure at moderate pressure of the coolant.

Explicit Calculation at Moderate Pressure

Especially at moderate pressure we can assume:

$$h'(T_{Leid}) - h'(T_g) = c_{pl}\,(T_{Leid} - T_g), \quad (21)$$

assumption will be emphasized when finally discussing the results (cf. fig. 2).

+) If the vapor g is assumed at a saturated state, the heat flux q_g^* (according to vapor without kinetic energy) cannot exist. Otherwise the vapor would become superheated with further conduction heat supply.

On the other hand the minimum heat flux q_w^* corresponding to the Leidenfrost phenomenon is about three orders less than the heat flux at departure from nucleate boiling. The Leidenfrost phenomenon occurs at minimum heat flux condition. This heat flux is not taken into account and we can assume the "heated" wall to be at adiabatic condition. With both, disapearing q_w^* and q_g^*, we have eq. (16).

Heat flux controlled measurements become unstable in the adjacent region of transition boiling and can be carried out at transient mode alone. On the other hand, temperature controlled measurements (e.g. Weber & Johannsen, 1989) are taken at steady state conditions and thus supported for a precision of measurements which emphasize assumption (16).

$$v'' >> v' \quad (22)$$

and

$$p_s\,(T_g)\; v''\,(T_g) = R/M \cdot T_g, \quad (23)$$

where R and M designate the ideal gas constant and the molar mass. Now defining the enthalpy of evaporation for so-called equilibrium processes:

$$h_{fg} = h''\,(T_g) - h'\,(T_g) \quad (24)$$

we get with (19):

$$(p_{Leid}/p_g) = 1 + 2\,M\,h_{fg}/(RT_g)$$
$$- 2\,M\,c_{pl}\,T_{Leid}/R\,(1/T_g - 1/T_{Leid}). \quad (25)$$

Using the Clausius-Clapeyron equation (5) we can substitute one of the expressions in parentheses to rewrite equation (25). Substituting the pressures we get:

$$T_{Leid,i} = (1-A\ln(2/A - B\,(T_{Leid,i-1} - T_g)/T_g + 1))^{-1}.T_g \quad (26)$$

and substituting the reciprocal temperatures we get:

$$(p_{Leid}/p_g)_i = 1 + 2/A$$
$$- AB\,\ln(p_{Leid}/p_g)_{i-1}/(1 - A\ln(p_{Leid}/p_g)_{i-1}), \quad (27)$$

where $A = R/M \cdot T_g/h_{fg}$ and $B = 2\,c_{pl}/(R/M)$;

$T_{Leid\infty} = T_{Leid}$ and $(P_{Leid}/P_g)_\infty = P_{Leid}/P_g$.

Using equations (26) or (27), the Leidenfrost temperature or the corresponding pressure of saturation can be calculated iteratively. Nevertheless the calculation is now explicit and especially if we start (index 0) with

$$T_{Leid,0} = T_g + 100\text{ K} \quad (28)$$

or

$$(p_{Leid}/p_g)_0 = 20, \quad (29)$$

the first approximation $T_{Leid,1}$ or $p_{Leid,1}$ will be sufficient to describe the Leidenfrost point for water at atmosheric pressure ($T_{Leid,1} \approx T_{Leid}$, $p_{Leid,1} \approx p_{Leid}$). This approxumation and the verification of equation (19) should be carried out by comparing the analytical results with documented measurements.

VERIFICATION BY EXPERIMENT

A first step to verify equation (19) by means of empirical results can be performed using the measurements of Feng & Johannsen (1989). A complete set of 349 documented wall temperature measurements each at the minimum heat flux condition is availlable from Mrs. Feng (1989). She carried out the measurements at various flow conditions using different inlet subcooling, mass flux and pressure. Looking at the high local equilibrium quality at the position where the Leidenfrost - phenomenon is measured, we assume spray cooling for most of the 349 documented measurements. Since equation (19) is assumed to describe the initial conditions for stable inverted annular flow only, we have to select comparable measurements.

Therefore we define the following procedure to find the most appropriate void condition at each pressure

1. The thermodynamic equilibrium quality:

$$X_{eq} = (\bar{h} - h')\,/\,h_{fg} \quad (30)$$

is assumed to give a realistic indication of the true quality X at near saturated boiling conditions alone. Therefore we choose the data set for the smallest inlet subcooling at a given pressure.

2. Redgardless of the mass flux we look for the measurement associated with the smallest thermodynamic equilibrium quality (30) since inverted annular flow will be the more probable the lower the quality is.

3. Some of the resulting measurements are reported with negative heat transfer (cooled solid surface),due to numerical problems when evaluating the two dimensional heat conduction within the test section. Those measurements are not taken into account.

Finnaly we result in T_{Leid} measurements at six different pressures (cf Table I). Thermodynamic properties of water are taken from Schmidt (1982) in order to compare these measurements with our calculation (19).

p_1 in MPa	T_{Leid} in K(°C) measured(Feng,1989)	T_{Leid} in K(°C) calculated,eq.(19)	X_{eq}	α_{eq}
0.11	504 (231)	494 (221)	0.0120	0.947
0.25	533 (260)	533 (260)	0.0182	0.926
0.40	558 (285)	557 (284)	0.0128	0.847
0.70	589 (316)	587 (314)	0.0026	0.288
1.00	598 (325)	607 (334)	0.0246	0.813
1.20	673 (400)	617 (344)	0.0404	0.858

Table I Comparison of calculated and documented Leidenfrost temperatures

Introducing the thermodynamic equilibrium void fraction

$$\alpha_{eq} = (1 + S\,.\rho_g/\rho_1 \cdot (1-X_{eq})/X_{eq})^{-1} \quad (31)$$

and ignoring the slip between vapor and liquid axial velocity

$$S = w_{ag}\,/\,w_{al} = 1 \quad (32)$$

we could consider that on the basis of table I alone the measurement at $p_1 = 0.7$ MPa probably corresponds to inverted annular flow conditions. The deviation between measurement and calculation is 2K where Mrs. Feng (1989) also assumed the uncertainty of her measurements to be in the range of ± 2 K.

With that information we are surprised that our calculation approximately holds (± 10 K) for the whole of table I except the measurement at $p_1 = 1.2$ MPa. Note that the equilibrium void fraction (31) does not give all the information about the flow regime that could be expected since assumption (32) is doubtful. The true void fraction is reduced by higher slip eq. (31). Thus more of the measurements shown in Table I could correspond to inverted annular flow.

Nevertheless the whole test run at 1.2 MPa was carried out at high inlet subcooling (30 K). Therefore it could occur that the measurement at $p_1 = 1.2$ MPa corresponds to the highest void fraction, cf statement 1, eq. (30).

At the pressure p_l = 0.11 MPa we could use eq.(26) and start with eq. (28). Then the first approximation $T_{Leid;1}$ = 494 K equals the exact result of eq. (19) up to the decimal point. We find the same conclusion for T_s ($p_{Leid;1}$) using eq. (27) from the start at eq. (29).

Looking for further verifications of our model we found documented measurements by Yao & Henry (1978) which where carried out for plated and unplated surfaces. Since Weber (1989) reported that there are considerable problems to be solved when calibrating temperatures at plated heat transfer surfaces, we only take into account those measurements which are reported for unplated heat transfer surfaces.

Figure 2 shows a comparison of our calculation with documented flow boiling measurements of water by Feng (1989). These are reported for a monel surface. Yao & Henry (1978) documented pool-boiling of water for a stainless steel surface. Additionally the true Leidenfrost measurements (dropets on a heated plate from molybdenum at subatmospheric pressure) by Testa & Nicotra (1986) are represented in this figure. We calculated equation (19) using properties of water from Schmidt (1982). The comparison shows close agreement between our model and the documented measurements which seem to be neither affected by the material (monel, stainless steel, molybdenum) from which the heat transfer surface is constructed nor by the experimental condition (i.e. flow boiling, pool boiling or droplets on a heated plate).

Indeed the measurements for pool boiling by Yao & Henry (1978) scatter considerably. That seems to be affected by the unstable location of the Leidenfrost point for the multidimensional heat transfer problem of pool boiling. Additionally the measurements of Yao & Henry (1978) deviate considerably from our calculation using eq. (19) at subatmosheric pressure. This could be explained by the hydrostatic pressure effect of the elevaled pool when considerable void fraction is produced.

Nevertheless, the maximum difference between the measurements of Feng (1989) and Testa & Nicotra (1986) and our calculation is ± 10 K. Most of the measurements deviate from our calculations by less than ± 3 K.

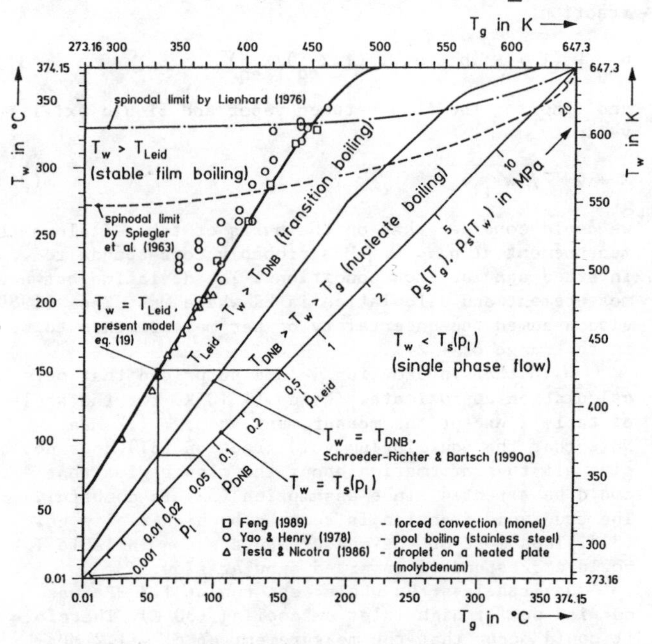

Fig. 2 Effect of pressure on the Leidenfrost temperature of water

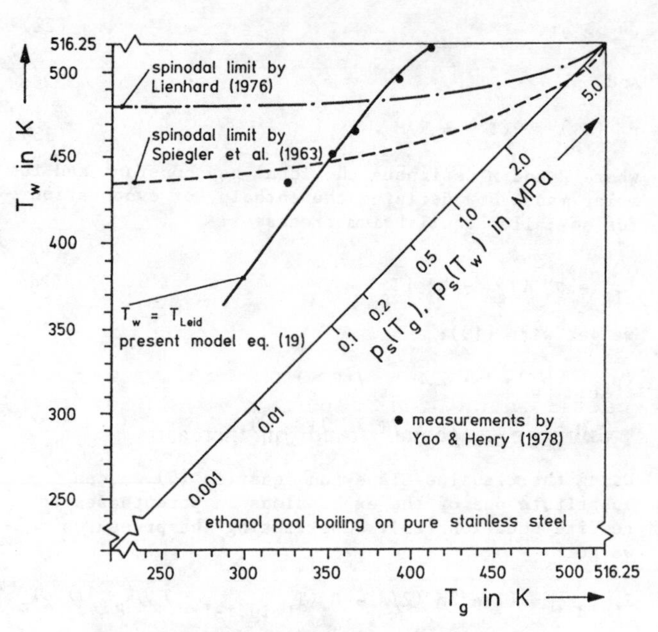

Fig.3 Effect of pressure on the Leidenfrost temperature for ethanol

Using the expression of Spiegler et al. (1963) the Leidenfrost temperature is as poorly estimated as if we had used the correlation of Lienhard (1976).

The same conclusion could be drawn when comparing the Spiegler et al. (1963) model and the Lienhard (1976) model with the experimental results for ethanol by Yao & Henry (1978),(cf fig.3). Our thermo-mechanical model on the other hand (eq. (19)) predicts the measured Leidenfrost temperatures well. Properties of ethanol are taken from Vargaftik (1975).

Fluid	saturation temperature T_g(K)	T_{Leid}(K) measured surface and geometry	T_{Leid}(K) calculated eq. (19)	$T_{Leid,\infty}$(K) calculated eq.(26)	author of the measurement and fluid properties
nitrogen	77	106 stainless steel pool boiling at standard and near zero gravity	108		Merte and Clark (1962) Sychev et al.(1987)
freon 10	350	432 aluminum pool/plate	458		Kautzky and Westwater (1967) Altumin et al.(1987)
freon 11	296	375 stainless steel pool/rod	352		Stock (1960) Altumin et al.(1987)
freon 113	320	399 aluminum pool/plate	properties of state not sufficient to solve eq.(19)	403	Kautzky and Westwater (1967) Vargaftik (1975)
n-pentane	309	368 copper pool/plate	properties of state not sufficient to solve eq.(19)	396	Berenson (1961) Vargaftik (1975)
mercury	630	843 tantalum drop	863		Poppendiek (1970) Vargaftik (1975)
sodium	1154	1594 tantalum pool/sphere	1595		Farahat et al.(1974) Vargaftik(1975)

Table II Measured and calculated Leidenfrost temperatures at atmospheric pressure

Leidenfrost temperatures at atmospheric pressure have recently been measured with various fluids and a collection of those measurements from cryogenic liquids up to liquid metals is given in table II. Using our thermo-mechanical model (eq. (19)) all measurements are well described in the range of experimental uncertainties.

Gunnerson & Cronenberg (1978) adapted the spinodal model using special equations of state for liquid metals. On the other hand the comparison in table II shows that no deviation at all could be observed when calculating the Leidenfrost temperature using our thermo-mechanical model (eq. (19)). Therefore our model seems to be more general for various fluids, different geometries of the boiling process (i.e. flow boiling, pool boiling, droplets on a heated plate) and for different materials from which the heat transfer surface could be made.

SUMMARY

The solid is partially wetted during transition boiling even though the wall temperature and thereby the temperature of the wetting liquid is considerably higher than the temperature of saturation at the corresponding pressure in the fluid center. It has previously been assumed that the pressure within the fluid is constant. On the other hand a liquid state in contact with the solid can only be explained with the additional assumption of a meta-stable region within the diagram of state. By using the limits of meta-stable states, thermodynamic models by Spiegler et al. (1963), Lienhard & Schrock (1966), Baumeister & Simon (1973) and others describe the temperature of the solid corresponding to the occurance of the Leidenfrost phenomenon. These limits are estimated at the local minima of isothermal pressure versus specific volume functions with the help of special equations of state (Van der Waals-EOS and others).

In contrast, hydrodynamic models by Berenson(1961), Gunnerson & Cronenberg (1980) and others make use of the Zuber (1958) analysis of the Helmholts instability, which results in a minimum heat flux corresponding to the Leidenfrost phenomenon. The Leidenfrost temperature is estimated by using special correlations of heat transfer.

However meta-stable models conflict with the results by van Hove (1949) and Prigogine & Defay (1950). Therefore a fundamentally new approach has been adopted defining both liquid and vapor to be in satured states at different pressures. The mechanical non-equilibrium is explained by the occurance of a thermo-mechanical effect. This model has recently been successfully applied to calculate nucleate boiling up to DNB (Schroeder-Richter & Bartsch 1990a).

In the present investigation we postulate that this thermo-mechanical effect holds during transition boiling. Assuming that the wetting liquid at the heated surface evaporates promptly when reaching the Leidenfrost temperature, the heated surface can be assumed to be adiabatic. Therefore the enthalpy of evaporation has to be provided from the mechanical energy alone when the wetting liquid is depressurized. The balance of energy now compares the properties of two saturated states alone. Thus the Leidenfrost temperature is well defined with the given system pressure for the center of the fluid and we calculate the Leidenfrost temperature analytically from a table of saturated states without using any empirical coefficient.

A comparison of our calculations with different measurements from cryogenic liquids, alcohol and water up to liquid metals shows excellent agreement. This is true for the different geometries of the boiling process (i.e. droplets on a heated plate, pool boiling, forced convection). Since a succesfull prediction of the Leidenfrost temperature is valuable, it has been usefull to assume that the thermo-mechanical effect holds during transition boiling.

Acknoledgement - The authors are very grateful to Prof. Johannsen and his colleages for their spontanous allowance to use unpublished measurements. Only the precision of their measurements encouraged us to follow up this idea with further investigations. The sudden and unexpected recent death of Professor Johannsen is of great sadness to the authors, who have lost a dear colleage and friend.

REFERENCES

Alamgir, Md., Kan, C.Y., and Lienhard, J.H., 1980, "An Experimental Study of the Rapid Depressurization of Hot Water", ASME Journal of Heat Transfer, Vol. 102, pp. 433-438.

Altumin, V.V., Geller, V.Z., Kremenevskaya, E.A., Perelshtein, I.I., and Petrov, E.K., 1987, "Thermophysical Properties of Freons: Methane Series, Part 2", National Standard Reference Data Service of the USSR: A Series of Property Tables, Vol. 9, Engl. ed. T.B. Selover, Jr., Hemisphere, Washington-New York-London.

Bartsch, G., and Schroeder-Richter, D., 1989, "Theoretical Investigation of the Mechanical Non-Equilibrium Caused by the Radial Temperature Distribution within Two-Phase Flows", Proceedings 7th Eurotherm Seminar, "Thermal Non-Equilibrium in Two-Phase Flow", ENEA, Rome, pp. 373-388.

Baumeister, K.J., and Simon, F.F., 1973, "Leidenfrost Temperature - Its Correlation for Liquid Metals, Cryogens, Hydrocarbons, and Water", ASME,Journal of Heat Transfer, Vol. 95, pp. 166-173.

Berenson, P.J., 1961, "Film-Boiling Heat Transfer From a Horizontal Surface", ASME Journal of Heat Transfer Vol. 83, pp. 351-358.

Biney, P.O., Dong, W.-g., and Lienhard, J.H., 1986, "Use of a Cubic Equation to Predict Surface Tension and Spinodal Limits", ASME Journal of Heat Transfer, Vol. 108, pp. 405-410.

Bjornard, T.A., and Griffith, P., 1977, "PWR Blowdown Heat Transfer", Proceedings, Thermal and Hydraulic Aspects of Nuclear Reactor Safety, O.C. Jones, Jr. and S.G. Bankoff, ed., ASME, New York, Vol. 1, pp. 17-41.

Blander, M., and Katz, J.L., 1975, "Bubble Nucleation in Liquids", AIChE Journal, Vol. 21, pp. 833-848.

Carbajo, J.J., 1985, "A Study on the Rewetting Temperature", Nuclear Engineering and Design, Vol. 84, pp. 21-52.

Carlslaw, H., and Jaeger, J., 1959, "Conduction of Heat in Solids", 2nd. ed., Oxford Clarendon Press, London.

Farahat, M.M.K., Eggen, D.T., and Armstrong, D.R., 1974, "Pool Boiling in Subcooled Sodium at Atmospheric Pressure", Nuclear Science and Engineering, Vol. 53, pp. 240-254.

Feng, Q., and Johannsen, K., 1989, "Minimum Heat Flux Condition in Convection Boiling Heat Transfer of Water", Proceedings International Conference on Mechanics of Two-Phase Flow, Taipei, pp. 463-469.

Feng, Q., 1989, Private Communication, Technical University of Berlin.

Førland, K.S., Førland, T., and Ratkje, S.K., 1988, "Irreversible Thermodynamics, Theory and Applications", John Wiley & Sons, Chichester-New York-Brisbane-Toronto-Singapore.

Gunnerson, F.S., and Cronenberg, A.W., 1978, "On the Thermodynamic Superheat Limit for Liquid Metals and Its Relation to the Leidenfrost Temperature", ASME Journal of Heat Transfer, Vol. 100, pp. 734-737.

Gunnerson, F.S., and Cronenberg, A.W., 1980, "On the Minimum Film Boiling Conditions for Spherical Geometries", ASME Journal of Heat Transfer, Vol. 102, pp. 335-341.

Himpan, J., 1955, "Die definitive Form der neuen thermischen Zustandsgleichung nebst ihren Stoffkonstanten von über 100 verschiedenen Stoffen", Monatshefte für Chemie und verwandte Teile anderer Wissenschaften, Vol. 86, pp. 259-268.

Hove, L. van, 1949, "Quelques propriétés générales de l'intégrale de configuration d'un système de particules avec interaction", Physica, Vol. 15, pp. 951-961.

Hove, L. van, 1950, "Sur l'intégrale de configuration pour les systèmes de particules à une dimension", Physica, Vol. 16, pp. 137-143.

Juza, J., 1965, "An Equation of State for Water and Steam", Proceedings, 3rd Conference Advances in Thermophysics, 1 Properites at Extreme Temperatures and Pressures, pp. 55-67.

Kautzky, D.E., and Westwater, J.W., 1967, "Film Boiling of a Mixture on a Horizontal Plate", International Journal of Heat and Mass Transfer, Vol. 10, pp. 253-256.

Leidenfrost, J.G., 1756, " De Aquae Communis Nonullis Qualitatibus Tractus", Duisburg on Rhine, C.Wares, 1966, tr., International Journal of Heat and Mass Transfer, Vol. 9, pp. 1153-1166.

Lienhard, J.H., and Schrock, V.E., 1966, "Generalized Displacement of the Nucleate Boiling Heat-Flux Curve, With Pressure Change", International Journal of Heat and Mass Transfer, Vol. 9, pp. 355-363.

Lienhard, J.H., 1976, "Correlation for the Limiting Liquid Superheat", Chemical Engineering Science, Vol. 31, pp. 847-849.

Lienhard, J.H., Alamgir, Md., and Trela, M., 1978, "Early Response of Hot Water to Sudden Relaese from High Pressure", ASME Journal of Heat Transfer, Vol. 100, pp. 473-479.

Lienhard, J.H., and Karimi, A.H., 1978, "Corresponding States Correlations of the Extreme Liquid Superheat and Vapor Subcooling", ASME Journal of Heat Transfer, Vol. 100, pp. 492-495.

Merte, H. & Clark, S.A., 1961, "Boiling Heat-Transfer Data for Liquid Nitrogen at Standard and Near-Zero Gravity", Advances in Cryogenic Engineering, Vol. 7 (1962), pp. 546-550.

Meyer, J., 1911, "Zur Kenntnis des negativen Druckes in Flüssigkeiten", Zeitschrift für Elektrochemie, Vol. 17, pp. 743-745.

Müller, I., 1985, "Thermodynamics" Pitman Advanced Publishing Program, Boston-London-Melbourne.

Muschik, W., and Müller, W.H., 1983, "Bilanzgleichungen offener mehrkomponentiger Systeme, part I and II", Journal of Non-Equilibrium Thermodynamics, Vol. 8, pp. 29-66.

Nishio, S., 1987, "Prediction Technique for Minimum-Heat-Flux (MHF) - Point Condition of Saturated Pool Boiling", International Journal of Heat and Mass Transfer, Vol. 30, pp. 2045-2057.

Poppendiek, H.F., 1970, "SNAP-8 Boiler Performance Degradation and Two-Phase Flow Heat and Momentum Transfer Models", GLR-84, NASA CR-72759, cited by Baumeister & Simon (1973).

Prigogine, I., 1947, "Étude thermodynamique des phénomènes irreversibles", Ph. D. Thesis, Univ. Bruxelles, ed., Dunod: Paris and Desoer: Liége.

Prigogine, I., and Defay, R., 1950, "Thermodynamique chimique", Desoer, Liége, Eng. tr. D.H. Everett, 1954, "Chemical Thermodynamics", Germ. tr. M. Winiker, 1962, Chemische Thermodynamik", VEB Grundstoffindustrie, Leipzig.

Schmidt, E., 1982, "Properties of Water and Superheated Steam in SI-Units, Zustandsgrößen von Wasser und Wasserdampf in SI-Einheiten, 0-800 °C, 0-1000 bar", 3rd enlarged printing, U. Grigull ed., Springer, Berlin.

Schroeder-Richter, D., and Bartsch, G., 1990a, "Analytical Calculation of the DNB-Superheating by a Postulated Thermo-Mechanical Effect of Nucleate Boiling", International Journal of Multiphase Flow, under review.

Schroeder-Richter, D., and Bartsch, G., 1990b, "The Growth of a Single Vapor Bubble Within an Isothermal Liquid - A Contribution to the Thermo-Mechanical Effect of Boiling", Journal of Non-Equilibrium Thermodynamics, under review.

Schroeder-Richter, D., and Bartsch, G., 1990c, "A New Model Describing the Superheated State of the Wall Layer During Nucleate Boiling", International Communications in Heat and Mass Transfer, Vol. 17, pp. 1-8.

Skripov, V.P., 1972, "Metastable Liquids", tr. from Russian R. Kondor, ed. D. Slutzkin, John Wiley & Sons, New York-Toronto, Israel Program for Scientific Translations, Jerusalem-London (1974).

Spiegler, P., Hopenfeld, J., Silberberg, M., Bumpus Jr., C.F., and Norman, A., 1963, "Onset of Stable Film Boiling and the Foam Limit", International Journal of Heat and Mass Transfer, Vol. 6, pp. 987-989.

Stock, B.J., 1960, "Observations on Transition Boiling Heat Transfer Phenomena", ANL - 6175, cited by Baumeister & Simon (1973).

Sychev, V.V., Vasserman, A.A., Kozlov, A.D., Spiridonov, G.A., and Tsymarny, V.A., 1987, "Thermodynamic Properties of Nitrogen", National Standard Reference Data Service of the USSR: A Series of Property Tables, Vol. 2, Eng. ed. T.B. Selover Jr., Hemisphere, Washington-New York-London.

Testa, P., and Nicotra, L., 1986, "Influence of Pressure on the Leidenfrost Temperature and on Extracted Heat Fluxes in the Transient Mode and Low Pressure", ASME Journal of Heat Transfer, Vol. 108, pp. 916-921.

Thom, J.R.S., Walker, M.W., Fallon, T.A., and Reising, G.F.S., 1965-66, "Boiling in Subcooled Water During Flow in Tubes and Annuli", Proceedings of the Institution of Mechanical Engineers, Vol. 180, No. 3C, pp. 226-246.

Vargaftik, N.B., 1975, "Tables on the Thermophysical Properties of Liquids and Gases", Hemisphere, Washington-London.

Waals, J.D. van der, 1873, "Over de continuiteit van den gas en vloeistof toestand, "Ph. D. thesis, Univ. Leyden, 2nd Germ. ed., Theil 1: 1899, Theil 2: 1900, "Die Continuität des gasförmigen und flüssigen Zustandes", Barth, Leipzig.

Weber, P., and Johannsen, K., 1989, "Temperature-Controlled Measurement of Boiling Curves for Forced Upflow of Subcooled Water in a Circular Tube at 0.1 to 1.0 MPa", ed. R.K. Shah, "Heat Transfer Equipment Fundamentals, Design, Applications, and Operating Problems, ASME HDT - Vol. 108, New York, pp. 13-21.

Weber, P., 1989, Private Communication, Technical University of Berlin.

Yao, S.-c., and Henry, R.E., 1978, "An Investigation of the Minimum Film Boiling Temperature on Horizontal Surfaces", ASME Journal of Heat Transfer, Vol. 100, pp. 260-267.

Zuber, N., 1958, "On the Stability of Boiling Heat Transfer", Transactions of the ASME, Vol. 80, pp. 711-720.

A GENERALIZED EMPIRICAL CORRELATION FOR MAXIMUM WALL SUPERHEAT IN TRANSITION BOILING UNDER FORCED CONVECTIVE CONDITIONS

Q. Feng and K. Johannsen
Institut fur Energietechnik
Technische Universitat Berlin
Berlin, Federal Republic of Germany

ABSTRACT

From temperature-controlled experiments of forced convective transition boiling heat transfer during upflow of water in a circular tube, 343 data points of wall temperature at minimum heat flux were deduced. An accurate generalized correlation was derived that reflects the parametric effects of pressure (0.11 - 1.2 MPa), local equilibrium quality (- 0.04 - 0.8), mass flux (10 - 1000 kg/(m^{2}s)), and inlet subcooling (4 - 57 K) on the maximum transition boiling temperature and has the correct asymptotic trends for zero flow and saturated conditions. Obviously, the high temperature limit of flow transition boiling is not thermodynamically controlled but, apart from pressure, depends on further local parameters as well as the upstream history of the flow.

NOMENCLATURE

a, b, c	= coefficients
c_p	= specific heat at constant pressure, J/(kg ^{o}C)
D	= inner tube diameter, mm
G	= mass flux, kg/(m^{2}s)
h	= enthalpy, J /kg
h_{lg}	= latent heat of vaporization, J/kg
Ja	= pool boiling Jakob number (= $\rho_l c_{pl} (T_L - T_s)/(h_{lg} \rho_g)$)
k	= thermal conductivity, w/(m ^{o}C)
L	= heated length of tube, mm
Pe	= Péclet number
p	= pressure, MPa
r	= linear correlation coefficient
T	= temperature, ^{o}C
T_L	= thermo-mechanical Leidenfrost temperature, ^{o}C
ΔT_L	= wall superheat at saturated pool boiling (= $T_L - T_s$), K
ΔT_{mtb}	= maximum wall superheat at transition boiling (= $T_{mtb} - T_s$), K
ΔT_s	= inlet subcooling (= $T_s - T_{in}$), K
x_e	= thermodynamic equilibrium quality
z	= distance form inlet, mm
ρ	= density at saturation, kg/m^3

Subcripts

g	= vapour or saturated vapour
in	= inlet
l	= liquid
min	= minimum film boiling
mtb	= maximum transition boiling
s	= saturation
w	= wall

INTRODUCTION

The limits of the transition boiling regime are conventionally defined by the maximum and minimum heat fluxes and the corresponding wall temperatures or wall superheats. During their experimental determination by means of power controlled experiments, one takes advantage of the fact that these region boundaries also form stability limits of either nucleate or film boiling. In contrast to the severe maximum heat flux limit (CHF), the minimum heat flux limit is much less pronounced. In convective boiling of water, for instance, the minimum of the boiling curve is typical very shallow and, over quite a range of wall superheat, the heat flux is almost constant as exhibited by recent experimental studies (Cheng et al., 1981, and Weber et al., 1988). Consequently, the empirical results obtained when approaching the minimum heat flux condition from film boiling are sensitive to how the experiment was performed. Thus various stability limits of film boiling according to different experimental situations are determined in terms of a surface temperature rather than the high-temperature limit of steady-state transition boiling. The corresponding conceptional difficulties are quite clearly reflected by the numerous terms that are presently in use to designate the various film boiling limits (e.g., Leidenfrost, calefaction, sputtering, apparent rewetting, true quench or minimum film boiling temperature among others).

Hence, Nelson's (1984) statement still applies that the current literature on the minimum surface superheat ΔT_{min} for forced convection film boiling provides, at best, a very confusing picture. The primary contributing factors in the confusion he quite correctly believed to be "(a) that the data reduction procedure may produce results dependent upon how the quenching transient, (b) that different types of experiments have been used, and (c) the effect of axial conduction upon ΔT_{min}". For a certain type of experiment, the remaining factors that may be responsible of the ambiguity can be removed by either reducing the measured transient temperature data via a solution of the appropriate multidimensional inverse heat conduction problem or performing a steady-state or, at least, quasi-steady experiment with subsequent data reduction through solution of the steady-state heat conduction problem.

Hence, performing surface temperature-controlled experiments to study flow boiling heat transfer in all regimes including transition boiling provides a means to resolve some of these difficulties. In this way it is possible to determine a unique high temperature limit of transition boiling that, by definition, corresponds to the minimum wall heat flux observed. In a recent experimental investigation of post-CHF heat transfer during forced upflow of water in a circular tube, we have applied temperature controlled indirect Joule heating which enabled us to also collect steady-state data for the such defined maximum wall superheat of transition boiling ΔT_{mtb}. Some of the results have been presented and discussed elsewhere (Feng and Johannsen, 1989). It is the purpose of this paper to present a generalized empirical correlation for ΔT_{mtb} that is primarily based upon these data but also includes recent results for higher pressure and mass flux.

EXPERIMENTS

Since the experimental system as well as the measurement and data evaluation procedures are described in detail elsewhere (Feng et al., 1987 - 1988, Feng and Johannsen, 1989, Weber and Johannsen, 1989), only a brief description is given here. The test section consists of a hollow copper cylinder of 50 mm length with an outer diameter of 32 mm and an inner bore of 10 mm. This cylinder may be heated by ten cylindrical heating elements of 3 mm O.D. that are located coaxially on a cirle of 11.5 mm radius (cf. Figure 1). It was soldered onto a Monel tube of 0.15 mm wall thickness which serves as the flow channel (I.D. 9.7 mm). The test section was vertically connected to a closed recirculating water loop. Degassed, duo-distilled, demineralized and microfiltrated water with an electrical conductivity of less than 10 µS/m was used as the test fluid. To secuer unconditional stable flow conditions throughout the experiments, an inlet throttling of approximately 0.3 MPa was applied. Tests performed with different upstream compressibilities (i.e., with plena of different volume) and with zero inlet throttling demonstrated the "stiff" character of the system as measured boiling curves remained unaffected by all these changes.

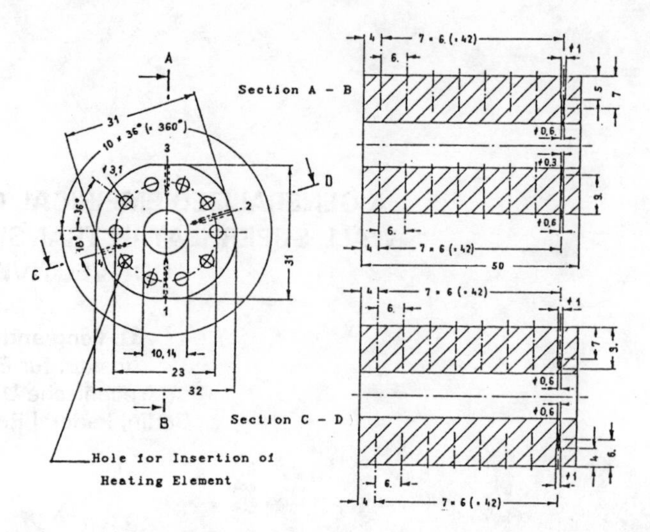

Figure 1. Copper test section

To realize measurements with prescribed (average) surface temperature, the electric heater power was controlled via electronic feedback control in dependence of this continuously sensed temperature as described previously (Johannsen and Kleen, 1984). In the present experiments the control point voltage was varied linearly with time corresponding to a time change of the reference temperature of 0.06 K/s. By comparison with steady-state experiments it has been proved that this rate of temperature change was securely small enough not to affect the stationary characteristics of the boiling process to be studied.

An on-line IBM-PC AT03 together with the Keithley Series 500 Data Acquisition and Control System was used for data acquisition and processing. The signals that are delivered from the test section instrumentation and the various water loop sensors were scanned every minute over differing time periods and averaged, thus constituting a data point.

The copper body of the test section is

instrumented by 32 microthermocouples to measure a two-dimensional temperature field in the r,z-plane pointwise. From these measurements, a continuous temperature field is constructed which yields the boundary conditions to a subsequent two-dimensional solution of the non-linear steady-state heat conduction equation governing heat transport within the Monel tube. In this way, the axial distributions of temperature and heat flux components at the heat transfer surface are obtained. Once these distributions are at hand, one may identify the minimum heat flux and its corresponding surface temperature in dependence of axial position by their judicious inspection (Feng and Johannsen, 1990).

From 55 measured runs, 343 data for the maximum transition boiling temperature were obtained. A complete tabulation of these data is being presented in a separate publication (Feng and Johannsen, 1990 a). The data cover the following ranges of independent flow parameters:

Pressure: $0.11 \leq p \leq 1.0$ (1.2) MPa

Mass flux: (10) $25 \leq G \leq 200$ (1000) kg/(m^{2}s)

Inlet subcooling: $4 \leq \Delta T_s \leq 30$ (57) K

Numbers in brackets indicate those parameter values that are not simultaneously met at their limits. The thermodynamic equilibrium quality x_e corresponding to T_{mtb} which will be negative for subcooled conditions ranges from - 0.04 to 0.8.

The uncertainty of the measured data was estimated by applying our previous uncertainty analysis (Feng et al., 1987 - 1988) to actually measured rather than "physically smooth" temperature signals (i.e., taking into account the additional periodic fluctuations introduced by the boiling process at minimum heat flux condition). Figure 2 shows the calculated 95 % confidence limits of the measured maximum transition boiling temperature and corresponding equilibrium quality for data of a characteristic run.

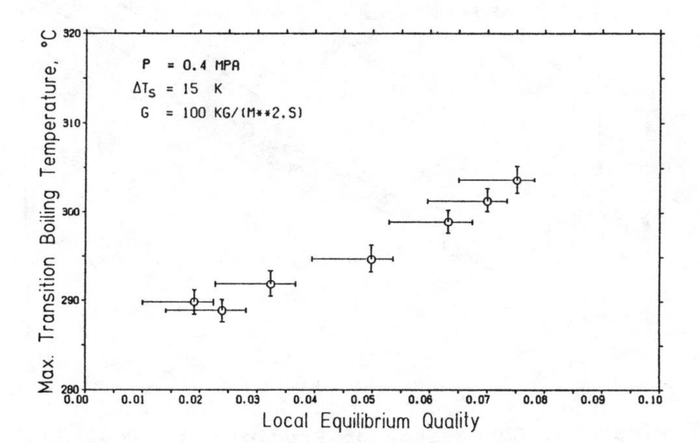

Figure 2. Measurement Uncertainty of Maximum Transition Boiling Temperature

RESULTS AND DISCUSSION

Dimensional Analysis

For correlating our data for the maximum trandition boiling temperature T_{mtb} in dependence of parameter variations we made use of dimensional analysis. Apart from the work by Kim and Lee (1979), no other reference could be found in the literature where this standard tool to facilitate and generalize the analytical presentation of experimental results had been applied to film boiling termination or rewetting during forced flow. As the variables that are assumed to affect the termination of transition boiling (i.e., T_{mtb}) we selected the tube dimensions (D, L), tube wall properties (c_{pw}, ρ_w, k_w), flow conditions at tube inlet (G, p, T_{in}) and flow parameters at locus (z) where T_{mtb} occurs (c_{pl}, ρ_l, k_l, h or q_{min}). For the pressure p we introduced the so-called "thermo-mechanical Leidenfrost temperature T_L" due to Schroeder-Richter and Bartsch (1990) which is a unique function of pressure. The temperatures T_{mtb}, T_L, and T_{in} as well as the enthalpy h(z) were replaced by their differences to saturation, that is, $\Delta T_s = T_s - T_{in}$, $\Delta T_{mtb} = T_{mtb} - T_s$, $\Delta T_L = T_L - T_s$, $\Delta h = h - h_{sl}$. These 14 variables must have a functional relationship

$$f(c_{pl}, c_{pw}, D, G, k_l, k_w, L, z, \Delta h, \Delta T_L, \Delta T_{mtb}, \Delta T_s, \rho_l, \rho_w) = 0. \quad (1)$$

Using Buchingham's Π-theorem with c_{pl}, k_l, ΔT_L and D as the primary variables and employing four primary dimensions of mass, length, time and temperature, ten non-dimensionless parameters Π_j (j = 10) are obtained. From these, the L/D ratio was removed as it was not varied in the experiments. In view of the calculated values of the linear correlation r_j for each Π-group left, four additional of these dimensionless parameters were found to affect the final result much less than the other ($r_j < 0.04$) and, hence, were neglected. The remaining Π-groups are the following

$$\Pi_1 = \frac{T_{mtb} - T_s}{T_L - T_s} = \frac{\Delta T_{mtb}}{\Delta T_L},$$

$$\Pi_2 = Re_{in}\, Pr_{in} = Pe_{in},$$

$$\Pi_3 = (x_e/Ja)(\rho_l/\rho_g),$$

$$\Pi_4 = \Delta T_s/\Delta T_L,$$

$$\Pi_5 = z/D.$$

The corresponding linear correlation coefficients r_2 through r_5 were: 0.776, 0.525, 0.125 and 0.047, respectively. Thus the data quite clearly indicate that T_{mtb} at forced convection is far from being thermodynamically controlled since it is severely affected by a number of other variables apart from pressure.

Correlation of Data

Correlation of data requires selection of a functional relationship of the normalized maximum wall superheat and the other dimensionless groups. Unlike to other heat transfer situations where an appropriate functional form is suggested by a basically correct analytical description of the governing physics, no much supporting information is available yet with regard to termination of convective transition boiling. Thus the choice of an adequate relationship had merely to be based upon certain asymptotic aspects that may reasonably be assumed or are exhibited by the data themselves (c.f., for example, the paper by Polyanin et al., 1989).

In the present case, it is reasonable to require that T_{mtb} approaches a finite limiting value characteristic for saturated free convective boiling (i.e., $G \to 0$ and $x \to 0$). Close inspection of the experimental data (Feng and Johannsen, 1990 b) reveals that, apparently, two different limiting values of transition boiling termination in saturated free convective boiling exist in accordance to the two main flow regimes that may occur upon an increase of wall temperature beyond T_{mtb} (inverted annular flow or dispersed flow). With regard to the z/D effect, T_{mtb} should approach a finite non-zero value when $z/D \to 0$; its importance is very likely to decrease with $z/D \to \infty$. Similarly reasonable assumptions regarding asymptotic trends of the other variables have also been made, sometimes upon suggestion by the behavior of the data if plotted in properly reduced variables. Such considerations ought to be reflected by the finally selected functional form of the correlation in order to secure that it will still yield acceptable predictions if applied beyond the parameter ranges of the underlying data base. Obviously, a simple power law relation as chosen by Kim and Lee (1979) for correlating the rewetting temperature does not satisfy this requirement.

The following expression was finally applied for correlating our data whose coefficients were found by Multiple Linear Regression Analysis:

$x_e < 0.1$:

$$\frac{\Delta T_{mtb}}{\Delta T_L} = a_o + a_1 (Pe_{in})^{a_2} (\Delta T_s/\Delta T_L)^{a_3} \{1 - a_4 \exp(-a_5 z/D)\} + \{b_1 - b_2 \exp(-b_3 Pe_{in}^{b_4})\}(x_e/Ja)(\rho_l/\rho_g) \quad (2)$$

$x_e \geq 0.1$:

$$\frac{\Delta T_{mtb}}{\Delta T_L} = c_o + c_1 Pe_{in}^{c_2} (\Delta T_s/\Delta T_L)^{c_3} \{(x_e/Ja)(\rho_l/\rho_g)\}^{c_4} \quad (3)$$

The corresponding coefficients are:

$a_o = 1.0$	$b_1 = 3.2$	$c_o = 0.65$
$a_1 = 0.102 \cdot 10^{-2}$	$b_2 = 3.363$	$c_1 = 0.102$
$a_2 = 0.859$	$b_3 = 0.31 \cdot 10^{-6}$	$c_2 = 0.306$
$a_3 = 0.830$	$b_4 = 1.5$	$c_3 = 0.792$
$a_4 = 0.986$		$c_4 = -0.264$
$a_5 = 0.20$		

The ranges of applicability of these Eqs. in terms of the dimensionless parameters are as follows:

Equation (2):

$1\,500 \leq Pe_{in} \leq 60\,500$

$0.03 \leq \Delta T_s/\Delta T_L \leq 0.4$

$-0.13 \leq (x_e/Ja)(\rho_l/\rho_g) \leq 0.45$

$2.89 \leq z/D \leq 4.4$

Equation (3):

$640 \leq Pe_{in} \leq 3\,200$

$0.04 \leq \Delta T_s/\Delta T_L \leq 0.35$

$0.4 \leq (x_e/Ja)(\rho_l/\rho_g) \leq 2.9$

The rms error of the predicted normalized wall superheat at termination of transition boiling as compared to our 343 measured data was calculated to be 9.3 %. The corresponding rms error of the maximum transition boiling temperature T_{mtb} was found to be 4.7 %. 98 % of the data are predicted within an error bound of ± 10 % as shown in Figure 3.

Discussion

Examination of Eqs. (2) and (3) reveals

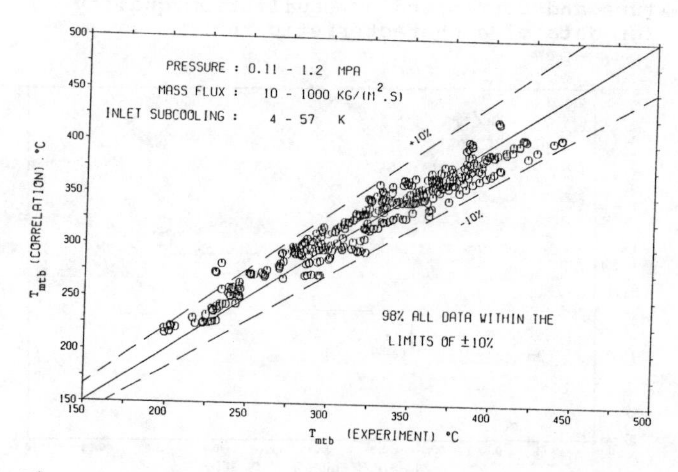

Figure 3. Comparison of Predicted and Measured Maximum Transition Boiling Temperature

that T_{mtb} approaches T_L for $x_e < 0.1$ and $0.65\ T_L$ for $x_e \geq 0.1$, respectively, if mass flux G and local equilibrium quality x_e simultaneously vanish ($x_e < 0.1$; Eq. (2)) or the mass flux is reduced till it reaches zero ($x_e \geq 0.1$; Eq. (3)). The distinction of the correlation according to quality with two differing free convective boiling limits was meant to take account of the apparently different flow situations from which this limit would be reached. At low void fractions, inverted annular flow film boiling will result upon an increase of the wall temperature beyond T_{mtb}, whereas dispersed flow film boiling will be established in case of higher void fractions.

As already mentioned above, we have selected the so-called "thermo-mechanical Leidenfrost temperature T_L" to be used as the free convective boiling limit in our correlation scheme, since it models termination of transition boiling which is just the subject of our study and, furthermore, its predictions are in excellent agreement with our data. According to Schroeder-Richter and Bartsch (1990) a thermo-mechanical effect is responsible for the liquid's loss of contact with the solid wall and the corresponding wall temperature may then be derived from purely thermodynamic considerations. This particular Leidenfrost temperature is unique function of pressure and, consequently, may be considered a thermodynamic property of the fluid. In view of the postulates underlying its derivation, T_L seems most appropriate to describe the critical transitional state at the boundary between low quality transition boiling and inverted annular flow film boiling at the zero flow and quality limit (saturated pool boiling). The following implicit expression for T_L is presented by the authors

$$h_g(T_g) - h_l(T_L) = \frac{1}{2}\{v_g(T_g) - v_l(T_L)\}\{p_g(T_L)-p_g(T_g)\} \quad (4)$$

where all variables refer to saturation at the temperatures indicated in round brackets ($T_L = T_W$). T_g is the saturation temperature of the vapour corresponding to the pressure in the liquid core (i.e., the pressure p of the system). For moderate pressure, an explicit approximation is also given in the reference.

For the case of high void fraction, if dispersed flow film boiling exists beyond T_{mtb} whose numerical value may be estimated by use of Eq. (3), we have tentatively chosen $0.65\ \Delta T_L$ as the free convective boiling limit. It remains an open question whether this value may be substantiated by future theoretical treatment of droplets' loss of contact with a hot solid wall due to a similar thermo-mechanical effect.

The adequate modeling of the pressure dependence of T_{mtb} data by incorporating T_L according to Eq. (4) into our correlation may probably be appreciated best when Figure 4 is consulted. In this figure, we have plotted sample predictions for the minimum film boil-

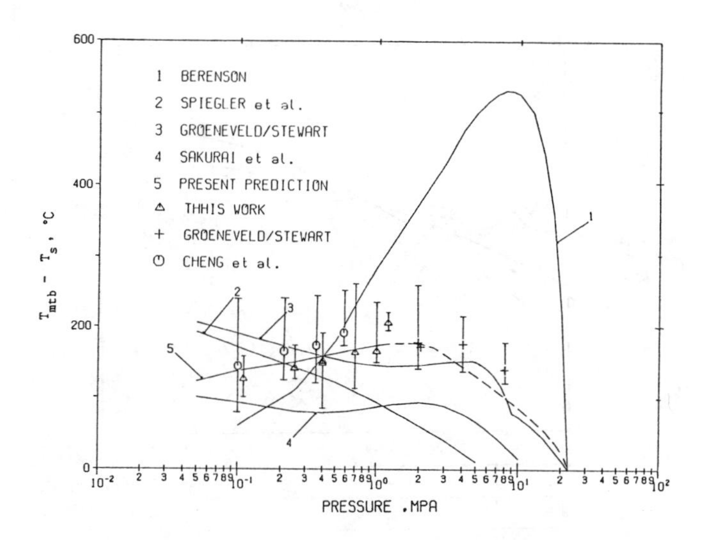

Figure 4. Effect of Pressure on Predicted and Measured Maximum Transition Boiling and Minimum Film Boiling Wall Superheat

ing temperature based on the maximum liquid superheat concept (Spiegler et al., 1963) and the Taylor-Helmholtz hydrodynamic instability theory (Berenson, 1961), respectively, together with a recent empirical pool boiling correlation for T_{min} on a horizontal cylinder (Sakurai et al., 1989) and the Groeneveld/Stewart correlation (1982) for T_{min} due to flow film boiling collapse. These predictions are compared with the present correlation, Eq. (2), at zero quality and a mass flux of $G = 100\ kg/(m^2s)$ and corresponding measurements. The bars attached to the measured points denote the measured variation of T_{mtb} to changes in mass flux that were unequal at different pressure. It is quite obvious that none of the other equations predicts the monotonous increase of the maximum wall superheat during transition boiling as exhibited by the measurements between 0.1 and 1 MPa, let alone the lack of any reasonable quantitative agreement.

Figure 5 exemplifies the sensitivity of normalized maximum transition boiling wall superheat, Π_1, to variations in the dimensionless parameters, Π_2 through Π_5, as exhibited by Eq. (2). The effect of z/D is least important; it almost disappears if $x_e \geq 0.1$. This fact is also elucidated by the numerical value of the linear correlation coefficient which was found to drop from 0.047 for the 249 low quality data to 0.027 for the 94 higher quality data. Consequently, the effect of z/D was omitted in Eq. (3).

To examen the functional dependencies of T_{mtb} on the independent parameters pressure and mass flux as well as local equilibrium quality as comprised in the correlation, we have produced some contour plots of predicted T_{mtb} with simultaneous variation of two parameters. For a fixed set of pressure, inlet subcooling and relative distance from inlet,

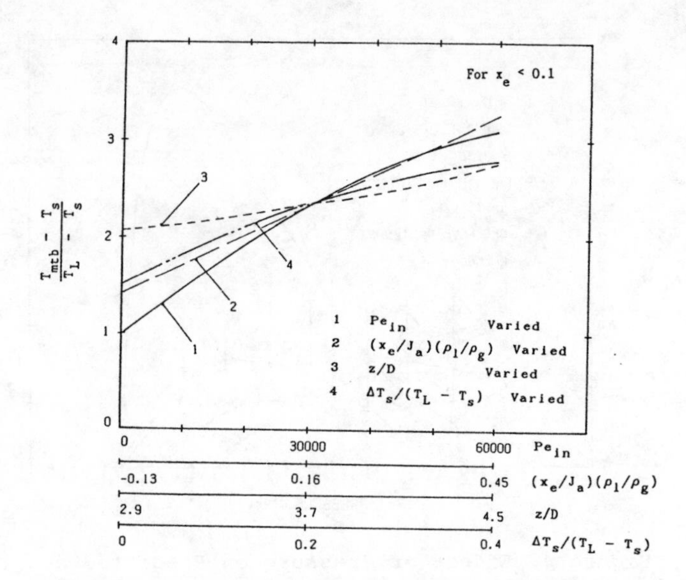

Figure 5. Sensitivity of Maximum Transition Boiling Temperature to Dimensionless Parameter Variations

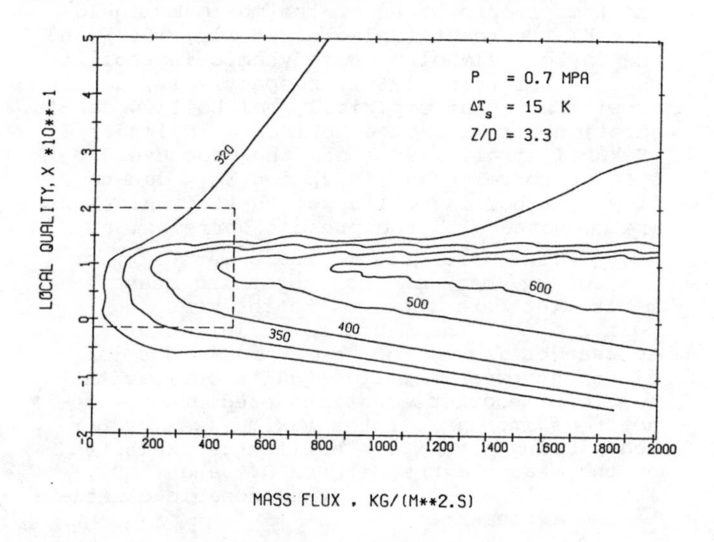

Figure 6. Contour Plot of Maximum Transition Boiling Temperature as Function of Local Equilibrium Quality and Mass Flux

isolines of T_{mtb} in the x_e,G-plane are plotted in Figure 6. For sake of a general overview, we tentatively applied the correlation also far beyond the range of data which is indicated by the rectangle. It seems that the mass flux effect tends to vanish as G increases regardless of quality. The effect of quality is much more significant except at very low mass flux or, in other words, close to pool boiling conditions. For a fixed mass flux, T_{mtb} generally increases with quality until $x_e \simeq 0.1$ and then decreases again. This change in trend, originally observed in the data (Feng and Johannsen, 1990 b), led to the formulation of two different correlating equations and the stated tentative quality limit.

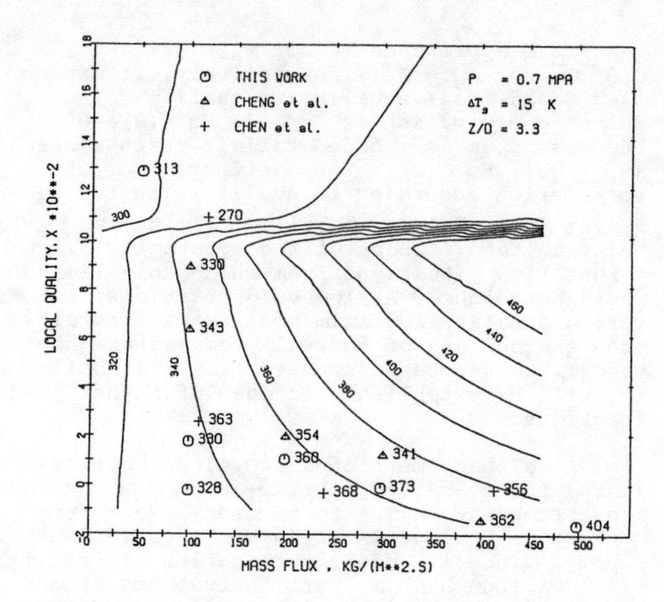

Figure 7. Contour Plot of Predicted Maximum Transition Boiling Temperature as Function of Local Equilibrium Quality and Mass Flux - Comparison with Experimental Data

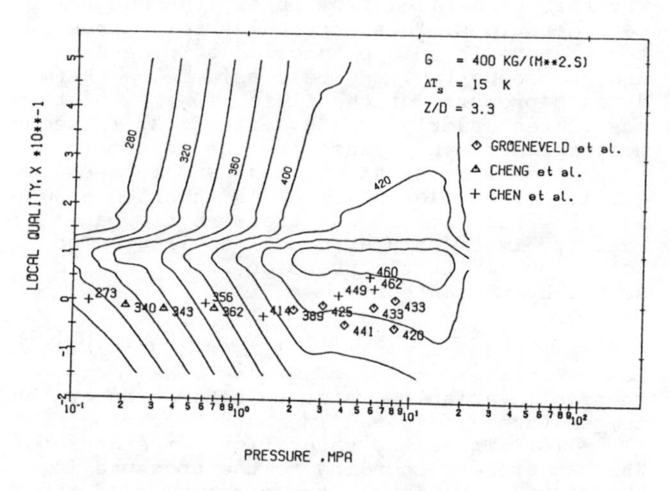

Figure 8. Contour Plot of Predicted Maximum Transition Boiling Temperature as Function of Local Equilibrium Quality and Pressure - Comparison with Experimental Data

It has to be admitted that distinction of different prevailing flow patterns as assumed by equilibrium quality is quite inadequate for obvious reasons but, unfortunately, no direct information on the flow field or, at least, the void fraction was available from the measurements. Hence, it would be very desirable to substantiate the present assumption on a change of flow pattern and the associated different mechanisms governing transition boiling termination in course of future measurements.

Both quality and mass flux are of similar importance on T_{mtb} at low and medium mass flux as demonstrated by Figure 7 where the parameter region covered by our data is enlarged form Fig. 6. In this figure, we have also plottet some experimental results for the "true quench temperature T_Q" due to impulse cooling collapse or dispersed flow rewet (Cheng et al., 1982, and Chen et al., 1989) for comparison. These data were obtained by approaching rewet during upflow of water in a circular tube under (almost) steady-state conditions and, hence, are thought to be insignificantly affected by axial conduction. Thus one might expect that this careful approach of the transition boiling regime from film boiling may yield roughly similar results as those from our reverse procedure. The comparison of these data with the prediction reveals an only gross agreement in the general trend of the quality effect whereas the mass flux effect seems to be somewhat masked by the unknown but typically large uncertainty in this type of experiment.

A similar comparison regarding the simultaneous effects of pressure and quality is presented in Figure 8. The agreement of our prediction with data from three different sources is quite reasonable, even far beyond its upper pressure limit of 1.2 MPa. The better agreement may be attributed to the well-known fact that experiments on film boiling heat transfer including film boiling termination may be performed with greater accuracy as mass flux and pressure increase.

CONCLUSION

Careful analysis and correlation of experimental data for the high temperature limit of flow transition boiling during upflow of water in a vertically mounted tube revealed strong indication of the occurence of different mechanisms that are responsible for transition boiling termination depending most probably on the flow pattern. Only in the limiting case of saturated pool boiling, the transition to film boiling may be considered as a purely thermodynamically controlled process which is adequately described by a new theory due to Schroeder-Richter and Bartsch (1990) involving the postulate of a governing thermo-mechanical effect.

ACKNOWLEDGEMENT

This study was performed under a contract between EURATOM and the Technische Universität Berlin (Research Contract No. 3013-86-07 EL ISP D) within the program Shared Cost Action - Reactor Safety 1985 - 87, Research Area No. 4: Study of Abnormal Behavior of LWR Cooling Systems.

REFERENCES

Berenson, J.P., 1961, "Film-Boiling Heat Transfer From a Horizontal Cylinder", Journal of Heat Transfer, Vol. 83, pp. 351-358.

Chen, Y., Wang, J., Yang, M., and Fu, X., 1989, "Experimental Measurement of the Minimum Film Boiling Temperature for Flowing Water", Paper presented at 2nd International Symposium on Multi-Phase Flow and Heat Transfer, Xi'an, China, Sept. 18-21.

Cheng, S.C., Poon, K.T., Ng, W.W., and Heng, K.T., 1981, "Transition Boiling Heat Transfer in Forced Vertical Flow", Final Rept. for the period June 1979 - June 1981, Argonne Natl. Lab., Contract nos. 31-109-38-3564 and 31-109-38-5503, University of Ottawa, Ottawa.

Cheng, S.C., Poon, K.T., Lau, P., and Ng, W.W.L., 1982, "Transition Boiling Heat Transfer in Forced Vertical Flow (Measurements of True Quench Temperature)", Final Rept. for the period June 1981 - June 1982, Argonne Natl. Lab., Contract no. 31-109-38-5503, University of Ottawa, Ottawa.

Feng, Q.-J., Johannsen, K., and Weber, P., 1987 - 1988, "Uncertainty Analysis for Numerical Modeling of Surface Heat Transfer Characteristics in Experiments - Application to Post-CHF Flow Boiling Tests", Experimental Heat Transfer, Vol. 1, pp. 277-297.

Feng, Q., and Johannsen, K., 1989, "Minimum Heat Flux Condition in Convective Boiling Heat Transfer of Water", Proceedings, International Conference on Mechanics of Two-Phase Flows, National Taiwan University, Taipei, R.O.C., pp. 463-469.

Feng, Q., and Johannsen, K., 1990 a, "Experimental Results of Maximum Transition Boiling Temperature During Upflow of Water in a Circular Tube at Medium Pressure", Experimental Thermal and Fluid Science, forthcoming.

Feng, Q., and Johannsen, K., 1990 b, "The High-Temperature Limit of the Transition Boiling Regime for Water in Vertical Upflow at Medium Pressure", Proceedings, 9th International Heat Transfer Conference, Hemisphere Publ. Corp., Washington, D.C., forthcoming.

Groeneveld, D.C., and Stewart, J.C., 1982, "The Minimum Film Boiling Temperature for Water During Film Boiling Collapse", Proceedings, 7th International Heat Transfer Conference, U. Grigull etal., ed., Hemisphere Publishing Corp., Washington, D.C., Vol. 4, pp. 393-398.

Johannsen, K., and Kleen, U., 1984, "Steady-State Measurement of Forced Convection Surface Boiling of Subcooled Water at and Beyond Maximum Heat Flux Via Indirect Joule Heating of a Test Section of High Thermal Conductance", Proceedings, 3rd Multi-Phase Flow and Heat Transfer Symposium-Workshop, Part B: Applications, T.N. Veziroglu et al., ed., Elsevier Science Publ., Amsterdam, pp. 755-776.

Kim, A.K., and Lee, Y., 1979, "A Correlation of Rewetting Temperature", Letters in Heat and Mass Transfer, Vol. 6, pp. 117-123.

Nelson, R.A., and Duffey, R.B., 1984, "Quenching Phenomena", Keynote Paper presented at International Workshop on Fundamental Aspects of Post-Dryout Heat Transfer, Salt Lake City, Utah.

Polyanin, A.D., Alvares-Suares, V.A., Dil'man, V.V., and Ryazantsev, Yu.S., 1989, "Experimental Data Processing by Means of 'Asymptotic Coordinates'", International Journal of Heat and Mass Transfer, Vol. 32, pp. 1401-1411.

Sakurai, A., Shiotsu, M., and Hata, K., 1989, "Effect of System Pressure on Minimum Film Boiling Temperature for Various Liquids", Proceedings, Fourth International Topical Meeting on Nuclear Reactor Thermal-Hydraulics, U. Müller et al., ed., Verlag G. Braun, Karlsruhe, Vol. 2, pp. 1092-1098.

Schroeder-Richter, D., and Bartsch, G., 1990, "The Leidenfrost Phenomenom Caused by a Thermo-Mechanical Effect of Transition Boiling - A Revisited Problem of Non-Equilibrium Thermodynamics", Paper submitted for presentation at 5th AIAA/ASME Thermophysics and Heat Transfer Conference, Seattle, WA, June 18 - 20.

Spiegler, P., Hopenfeld, J., Silberberg, M., Bumpus jr., C.F., and Norman, A., 1963, "Onset of Stable Film Boiling and the Foam Limit", International Journal of Heat and Mass Transfer, Vol. 6, pp. 987-989.

Weber, P., Feng, Q., and Johannsen, K., 1988, "Study of Convective Transition Boiling Heat Transfer in a Vertical Tube with Slightly Subcooled Water at Medium Pressure", Paper presented at the 5th Miami International Symposium of Multi-Phase Transport and Particulate Phenomena, 12 - 14 Dec., Miami Beach, Florida, USA.

Weber, P., and Johannsen, K., 1989, "Temperature-Controlled Measurement of Boiling Curves for Forced Upflow of Water in a Circular Tube at 0.1 to 1 MPa", in Heat Transfer Equipment Fundamentals, Design, Applications, and Operating Problems, R.K. Shah, ed., ASME HTD-Vol. 108, ASME, New York, NY, pp. 13-21.

HIGHLY SUBCOOLED FLOW BOILING OF FREON-113 OVER CYLINDERS

S. Sankaran and L. C. Witte
Department of Mechanical Engineering
University of Houston
Houston, Texas

ABSTRACT

The intent of this study was to determine the influence of high subcooling and velocity on the ratio of minimum to maximum heat flux. The configuration was upflowing Freon-113 in crossflow over 0.635-cm electric heaters. As a part of the study, azimuthal temperature variations during nucleate and film boiling were measured, the manner of onset of film boiling and vapor film collapse was observed and photographed, and the effects of subcooling and velocity on q_{min} and q_{max} were determined.

Experiments were conducted at subcooling levels up to 58 C and at velocities up to 3.81 m/sec. While both q_{min} and q_{max} were sensitive to velocity and subcooling, the q_{min} to q_{max} ratio was seen to be a stronger function of subcooling than velocity. Previous investigators have reported that the q_{min} to q_{max} ratio for Freon-113 increases from 0.083 for pool boiling to 0.303 for 6.8 m/sec at low subcooling levels. In our work at 57 C subcooling and 3.53 m/sec, the ratio was found to be 0.771.

Discontinuities in measured wall temperatures were observed when the vapor film collapsed over certain portions of the heater prior to the entire heater being enveloped in nucleate boiling. Our observations confirm such behavior previously observed in our laboratory. These discontinuities are caused by film and nucleate boiling coexisting in near proximity close to the location of thermocouples mounted to measure heater local temperatures.

It was also observed that the bottom of the heater underwent an unexpected temperature drop during nucleate boiling for high levels of subcooling. Such temperature drops were very repeatable. They resemble the "temperature overshoot" phenomena that have been observed in pool boiling experiments on a variety of heating surfaces and configurations. However, no observable significant increase in nucleation that is generally considered to be responsible for such behavior was found in our experiments.

NOMENCLATURE

q heat flux

q_{max} critical heat flux

q_{min} minimum heat flux

T_{bulk} liquid temperature

T_{sat} saturation temperature

T_w wall temperature

V liquid velocity

T_{sub} liquid subcooling, T_{sat} - T_{bulk}

T_w wall superheat, T_w - T_{sat}

θ angle measured from forward stagnation point

INTRODUCTION

High heat fluxes at high temperatures will be required in many energy systems being developed for the future. For example, in fusion reactors, the heat that is deposited in the first wall containment must be removed at a high flux level and at extremely high temperatures to be used in a power conversion cycle. Another example is the receiver wall in a solar receiver that is supplied extremely high heat fluxes by a field of heliostats. Many exothermic reactions also occur at high temperatures and must be cooled to maintain their stability. Even high speed computers have a need for high flux cooling of their electronic components, although at relativley low surface temperatures.

Boiling has long been recognized as one of the most efficient ways of removing heat from a heated surface. Enormous fluxes can be accommodated at very low temperature differences between the surface and the surrounding coolant. However, the boiling process has a major limitation; the heat flux reaches a peak at a relatively low temperature difference (at a correspondingly low surface surface temperature). This peak is called the burnout point because in electrically-heated systems, the heating surface usually melts as the boiling undergoes a rapid jump to film boiling. In systems where temperature difference is the controlling parameter, a jump does not occur, but the heat flux actually drops dramatically as the surface temperature is increased. This puts a natural limitation on using a boiling process for surfaces that, for one reason or another, want to operate at a temperature greater than the burnout point temperature.

There is evidence that the traditional pool boiling curve can be dramatically altered by using a highly subcooled liquid moving at high speed over the heated surface. The influence of velocity and subcooling is more dramatic in the film and transition region than it is in the nucleate boiling region. Nucleate boiling is not affected as much as film and transition because the intense bubble action is not very sensitive to motion or liquid temperature. In contrast, in the film boiling region where the fluxes are lowest, velocity and subcooling have a larger impact because the processes are controlled more by the convective phenomena in the over-riding liquid.

Stevens and Witte (1973) provided evidence of the dramatic enhancement of heat fluxes in the film and transition regions in studies of quenching spheres traversing through subcooled water. They observed a factor of ten increase in the minimum heat flux as the water temperature decreased from 77C to 24C at a sphere velocity of 5 ft/sec (1.52 m/sec). The corresponding peak heat fluxes increased by only a factor of two for these experiments. This leads one to believe that for sufficiently high subcooling and velocity levels the peak and minimum fluxes might approach each other and no unstable transition boiling region would exist. The data of Yilmaz and Westwater (1980) and Broussard and Westwater (1984) also support this notion.

The data for highly subcooled sodium of Witte (1967) showed that there was no dimunition of heat transfer in what would seem to be the transition region of boiling. Bradfield (1960) also showed that high speed and subcooling causes a breakdown of the film boiling layer at high temperature into a "frothy" boundary layer type of flow, reminiscent of nucleate boiling at much lower temperatures.

All these studies indicate that it might be possible to avert the transition boiling region if the liquid temperature is very low and the liquid speed is high. In what follows we describe the results of a study to investigate whether or not this is possible.

EXPERIMENTAL APPARATUS

Figure 1 shows a schematic layout of the flow loop designed and fabricated to produce steady, low temperature flows over an electric heater. Liquid velocity was controlled by a variable speed 450 gpm pump (1). The flow passed through various flow straightening sections (4-7) before passing through the transparent plexiglas test section (8) across which the heater was suspended. A water-cooled Filtrine POC-500 W of 16 kW cooling capacity was used to maintain the flowing coolant at the desired temperature. Temperatures down to 0 C were possible with the use of this chiller. The system was designed such that the temperature of the working fluid would rise less than 1.0 C/min at the maximum heating rates.

Flow rates which were used to calculate flow velocities in the test section were measured with an annubar element (3). A pitot tube traverse accross the test section was used to verify that the calibration curve of the annubar was accurate. The pitot traverse also showed that the velocity profile in the test section was basically flat with the exception of thin boundary layers developing along the walls. Thermocouples (18) and pressure taps were installed at various points in the loop to yield the operating conditions of the working fluid.

Figure 2 shows a cutaway view of the electric heater. It was 0.635 cm in diameter and consisted of a thin--walled Hastelloy-C outer heating shell (0.04 mm thickness) fitted over a machined lava insert. The lava insert provide for structural rigidity as well as additional thermal mass. It also served as the vehicle for situating five 38 gage chromel-alumel thermcouples at various strategic locations adjacent to the heating surface as shown in Figure 2. After grooves were machined into the lava insert, and the thermocouples positioned properly in it, then the insert was heated at 1100 C for 24 hours so that the lava expanded to create very good contact between all elements of the heater system.

The manner in which the heater was mounted through the test section wall is also shown in Figure 2. Two thermocouples were mounted on the bus bars outside the test section wall; they were used to estimate the end losses of the heaters during operation. The heat losses through the ends of the heater were less than 2% of the total heat dissipated even for the highest heat fluxes encountered.

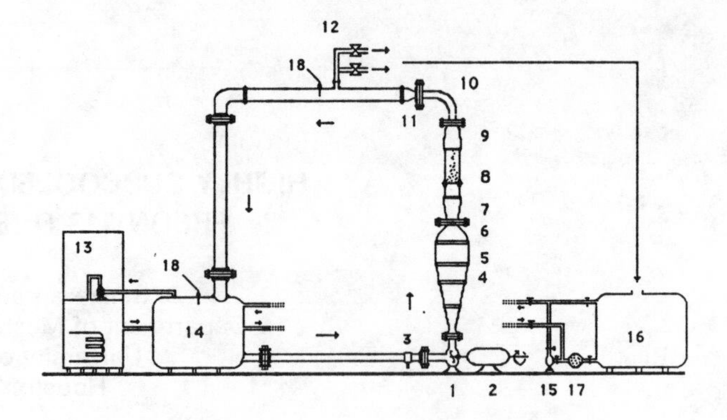

1.	Pump	11.	3" x 6" bell reducer
2.	25 hp motor	12.	Atm. vent, Pressure relief valve
3.	Annubar	13.	Chiller
4.	Diffuser	14.	55 gallon collection tank
5.	Honey comb	15.	1- hp pump
6.	Nozzle	16.	150 gallon storage tank
7,9.	Transition sections	17.	Filter
8.	Test section	18.	Thermocouples
10.	3" pipe bend		

Fig. 1. Schematic layout of the flow loop - chiller.

Power was supplied to the heater by a dc motor-generator set capable of providing up to 1500 amps of electric current at low voltages. Diesel locomotive cables (3/0, 450 wire strand) connected the heater to the motor-generator set. The energy being dissipated in the heater was found by simultaneous measurements of voltage and current provided to the heater.

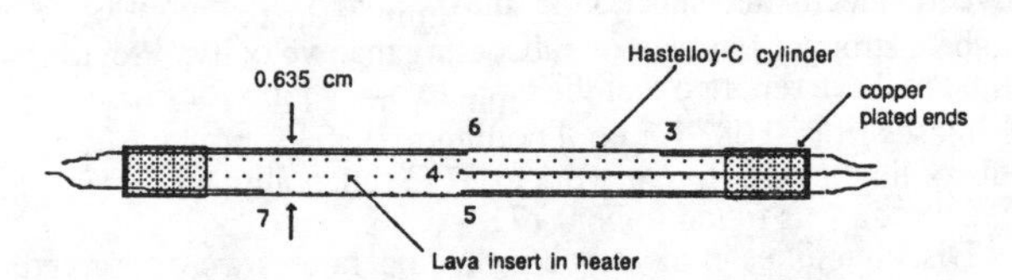

Fig 2.a Cutaway view of the electric heater

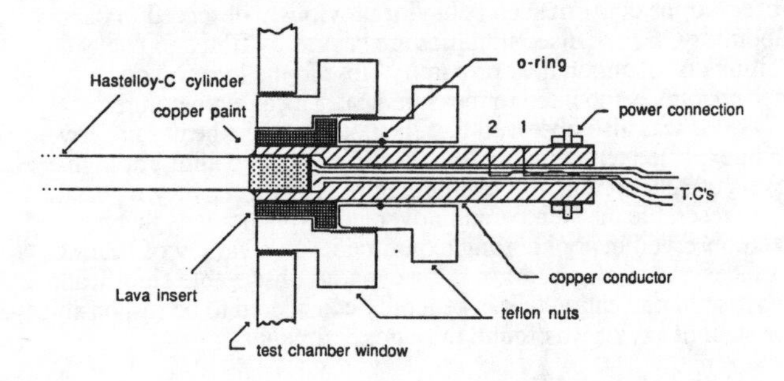

Fig 2.b Schematic showing the mounting of the heater in the test section.

1, 2. Outside thermocouples.
3, 7. End thermocouples.
4. Center thermocouple at 90 degree location.
5. Center thermocouple at the lower stagnation point.
6. Center thermocouple at the upper stagnation point.

EXPERIMENTAL CONDITIONS

Table 1 shows the range of conditions over which experiments were performed. It was found that the velocity was limited to about 3.8 m/sec with this apparatus. Above the flow rate corresponding to this velocity, large vibrations were imposed on the flow loop by the motor/controller combination.

TABLE 1

RANGE OF OPERATING CONDITIONS

Run No.	Velocity	T_{sat}	Subcooling
1	2.95 m/sec	60.0 C	50.0 C
2	2.95	61.0	51.0
3	2.95	59.5	54.5
4	3.53	65.5	58.6
5	3.81	68.1	57.6
6	3.53	65.5	57.0
7	2.95	60.0	56.5
8	2.95	60.0	55.6
9	2.95	59.5	49.3
10	2.95	61.0	42.0

An error analysis showed that heat flux could be determined to within $\pm$ 3%, while the temperature accuracy is estimated at $\pm$ 1.5 C in T_{bulk}, and < 2% in T_{heater}, which translate to $\pm$ 6 C at 300 C. Velocity in the test section could be determined to within $\pm$ 3 %.

VISUAL OBSERVATIONS OF THE BOILING PHENOMENA

A very important part of this study was to determine the flow patterns that occur when various boiling regions are encountered. This information can be used to gain insight into how the processes might be modelled. In addition to naked eye observations, a Spin Physics SP-2000 Motion Analyzer System was used for motion picture analysis while a Nikon still camera was used to capture various flow region information.

At very high subcoolings not much net vapor is produced around the heater, regardless of the boiling region. Bubbles quickly condense as they move away from the hot surface, while vapor film regions that occur during film boiling are kept very thin by the influence of subcooling.

Nucleate Boiling

At low heat fluxes, no boiling was present and non-boiling forced convection correlations compared well with the experimental data. As the heat flux was increased so that the surface temperatures increased well beyond the local saturation temperatures, nucleate boiling ensued. Bubbles were observed being formed on the surface and being swept around the periphery of the heater. A wake was formed, but it was a wake that consists primarily of liquid rather than primarily of vapor as is the case for nearly saturated boiling, see Lienhard and Eichhorn, 1976, and Ungar and Eichhorn, 1988. Figure 3 is a sketch of the heater that shows how bubbles were swept away from the cylinder surface near the 90-degree line on the heater. Bubbles could be observed forming in the wake region and being entrained in the rapidly flowing liquid at the edge of the wake. Higher heat fluxes increased the number of bubbles so that the wake-line (the point where bubbles are swept away from the surface) became much more visible.

As the heat flux was increased further in the nucleate boiling region, the bottom part of the cylinder consistently and repeatably underwent an unexpected temperature drop for the high subcooling experiments. But no visual change in either the wake pattern or the

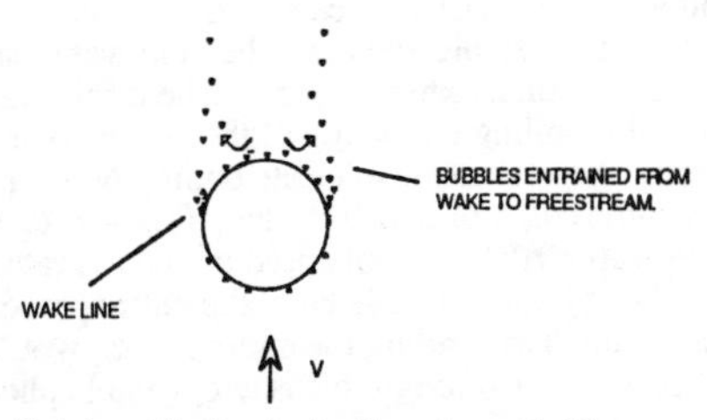

Fig. 3. End view of heater showing the pattern of bubbles being swept from the heater.

production of bubbles around the front of the heater was observed. This effect will discussed further in a later section.

Film Boiling

The first appearance of film boiling on the heater was somewhat random; in some cases it began on the ends of the heater and in other cases a small film boiling patch developed on the top of the cylinder as shown in Figure 4. As heat flux was increased slightly, film boiling would ensue on the top of the cylinder near the ends. These film boiling patches increased in size as heat flux was increased, eventually covering the entire heater surface, top and bottom. It was clear that for the conditions of these experiments,

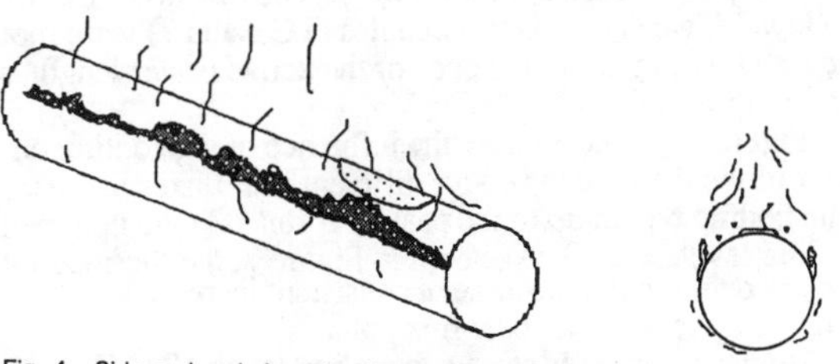

Fig. 4 . Side and end view showing how bubbles and a film boiling patch can coexist on the heater.

film and nucleate boiling could coexist on the heater surface. These regions were separated by transition "fronts"which take on a significantly different appearance than nucleate or film boiling. The region near the front is characterized by the production of many small bubbles that are rapidly diffused into the surrounding liquid.

Film boiling for all these high velocity, high subcooling experiments showed a much thinner wake, more like a tear drop, than the nucleate boiling wake, as shown in Figure 5. An oscillating separation line was observed slightly past the 90-degree point on the heater, also shown on Figure 5.

Once film boiling was established and the heat flux was decreased slowly, the vapor film would begin to collapse on the bottom of the heater which was colder than the top which was covered by the wake. The collapse was characterized by the disappearance of the liquid vapor interface at the lower stagnation point of the heater. A transition front as described previously separated film from nucleate boiling, and it was characterized by

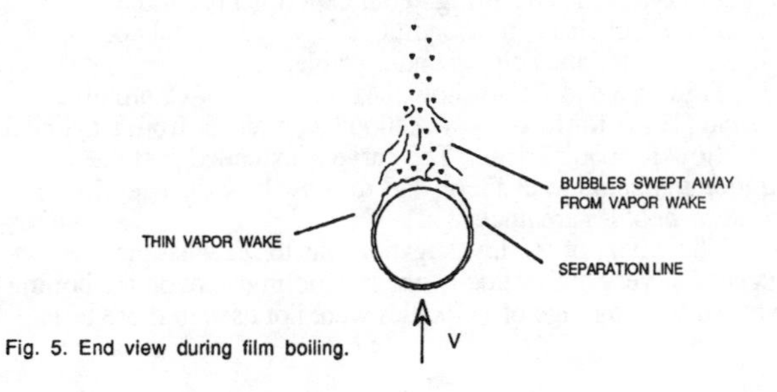

Fig. 5. End view during film boiling.

accentuated bubble activity as described before. The situation of film boiling on the top of the heater and nucleate boiling on the bottom was highly stable and could be (and was) maintained for several hours in some experiments. As heat flux was decreased further the film boiling on the top of the heater collapsed in a patchy manner leading to complete nucleate boiling on the heater.

In some cases in which the heat flux was decreased very rapidly, the vapor film first collapsed on the top rather than the bottom of the cylinder. In this case, the entire process of destruction of the vapor film surrounding the entire heater would occur in a matter of seconds. Although this is a repeatable phenomenon, it is as yet unexplained.

HEAT TRANSFER RESULTS

It is logical to assume that the bottom of the heater should always be colder than the top of the heater which is enveloped in a wake. The cold liquid comes into contact with heater at the lower stagnation point, and a thermal boundary layer of sorts develops as the liquid moves around the heater. While in general this was always the case for both film and nucleate boiling, some unexpected behavior of the temperatures was observed.

Nucleate Boiling

The effect of subcooling - The top and bottom thermocouples behaved differently as the subcooling was varied for a given velocity. At the center of the heater (midway across the test section) the bottom, middle and top thermocouples refer to the bottom stagnation point (TC 5 in Fig. 2a), the 90-degree point (TC 4) and the top stagnation point (TC 6) respectively in the discussion that follows. Two "end" thermocouples (TC 3 and 7) were mounted 0.5 inches (1.27 cm) from the ends of the active heater length, as shown in Figure 2a.

Figures 6, 7 and 8 show the influence of subcooling on the behavior of the thermocouples for different heat fluxes in the nucleate boiling region up to the peak heat flux. At the intermediate subcooling level, 49.5 C, as shown in Figure 7, the thermocouples behaved as one might expect; i.e, a consistent increase in temperature up to the peak heat flux point.

However, at the highest subcooling level, 57 C (Fig. 6), the bottom and middle thermocouples showed an unexpected decrease in temperature as the peak heat flux was approached. Such a decrease obviously means that the heat transfer coefficient over that thermocouple location must have increased dramatically with heat flux because of the constant heat flux condition at the surface. This could be explained by a dramatic increase in the population of bubbles at this transition point. However, as stated in the previous section, there was no visual evidence that such an increase in the bubble density occured at these transition points. Many tests for repeatabiity of this temperature behavior were made to insure the integrity of the measurement system. In all cases the data were repeatable. Such behavior has been observed in other nucleate boiling systems. Jung and Bergles (1989) observed this type of behavior for saturated Freon-113 boiling from plain, polished, electrically-heated copper tubes as well as from "enhanced" tubes, such as Gewa-T tubes. This behavior is attributed to the suppression of nucleate boiling up to wall temperatures typical of nucleate boiling, followed by a sudden onset of nucleate boiling as heat flux is increased. The literature related to cooling of electronic cooling by immersion boiling is also filled with references to the "temperature overshoot" problem, see for example, Anderson and Mudawwar (1988) and Bergles and Kim (1988). It is possible that such a phenomenon is occuring in our experiments but the population of bubbles at the conditions of very high subcooling is so small that changes are visually undetectable.

Figures 6 and 7 also show the correlation of Churchill and Bernstein (1977) for forced convection heat transfer from a cylinder in crossflow for comparison. The curve is extended past the saturation temperature of Freon-113 to show how the heat fluxes increase as bubbles are nucleated.

The intent of our investigation was to see what the influences of very high velocity and subcooling were on the boiling curves, so a wide range of velocities were not used in these tests.

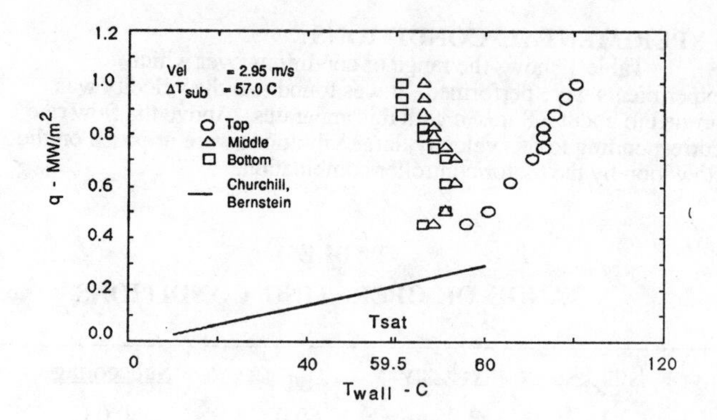

Fig. 6. Nucleate boiling curves for 57.0 C subcooled Freon-113.

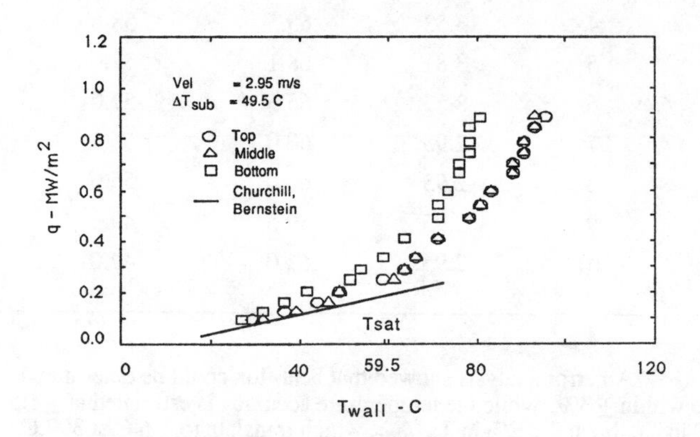

Fig. 7. Nucleate boiling curves for 49.5 C subcooled Freon-113.

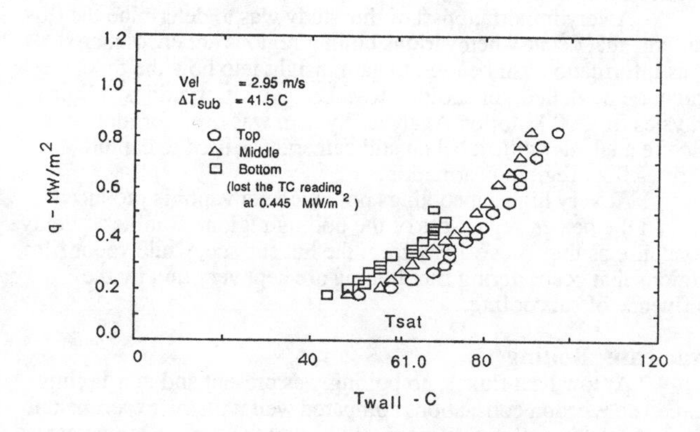

Fig. 8. Nucleate boiling curves for 41.5 C subcooled Freon-113.

Instead, we were interested in producing the largest velocity possible with our apparatus. For Freon-113, that limit was about 3.8 m/sec., as discussed previously.

Film Boiling

Azimuthal temperature differences were more pronounced in the film boiling region than in the nucleate boiling region. In film boiling, the top could be more than 115 C hotter than the bottom of the heater. Extensive experiments were carried out at 2.95 m/sec at different subcooling levels, and the effect of velocity was measured at a subcooling of 58 C.

Because the temperatures were so different between the top and the bottom of the heater in film boiling, a complete two-dimensional analysis of conduction within the heater, including heat transfer across the lava as well as around the periphery of the Hastelloy heater was performed. For the worst case that we computed, that is, for a temperature difference from top to bottom of 200 C, only a 6% deviation from a uniform heat flux condition was found, and this was at the very top of the heater. Other locations showed a much smaller deviation. In what follows, a uniform flux is assumed for the heater.

The Effect of Subcooling - Figures 9, 10, and 11 show the effect of subcooling at 2.95 m/sec. Near q_{min}, the temperature differences between the top and bottom decreased. As the heat flux was decreased slowly, the vapor film collapsed on the bottom and film boiling prevailed at the top as described previously. Thus, at the top the temperatures remained high typical of film boiling, while

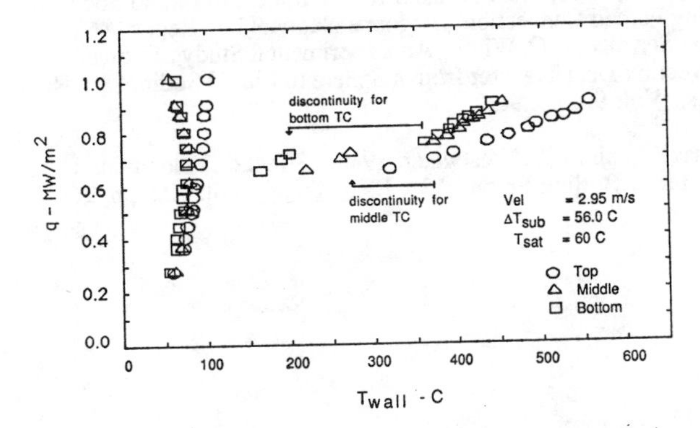

Fig .9 Boiling data showing nucleate and film boiling data for 58 C subcooled Freon-113.

the bottom and center did not immediately fall all the way into the range typical of nucleate boiling. Rather as seen in Figure 9, for example, the temperatures arranged themselves in decreasing temperature from hottest at the top to coldest at the bottom. Thus we see discontinuities in the bottom and middle thermocouple readings, as shown on Figure 9, prior to the entire surface being enveloped in nucleate boiling. The simultaneous existence of film and nucleate boiling on the surface affects the temperature that can exist at the bottom and middle thermocouples locations even though it is clear that film boiling no longer exists over those thermocouple locations. The middle thermocouple reads a higher temperature than the bottom one because it is closer to the film boiling patch at the top of the heater.

Chang and Witte, 1988, observed similar behavior for an electric heater in flowing Freon-11 using a single surface thermocouple designed to detect liquid-solid contact during film boiling. In their experiments, which were for very low levels of

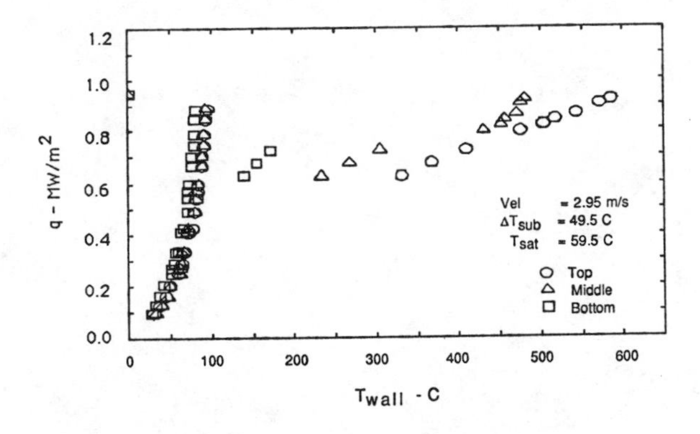

Fig. 10. Boiling data showing nucleate and film boiling data for 49.5 subcooled Freon-113. (nucleate boiling data same as Fig. 7).

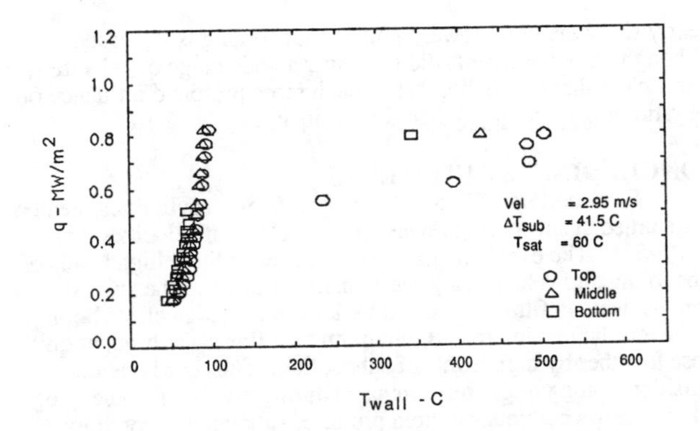

Fig . 11 Nucleate and film boiling data for 41.5 C subcooled Freon-113. (nucleate boiling same as Fig. 8)

subcooling, they observed curious intermediate jumps from clearly film boiling temperatures down to a much lower temperature but not down to the ultimate nucleate boiling temperature range as the heat flux was decreased. They speculated that this must be a stable operating point that could be supported because of the existence of different boiling regions on the surface of the heater. The present results now show more clearly that this is indeed what is happening.

The existence of different regions of boiling in close proximity on a surface can produce discontinuities in temperature at a particular location as the heater goes through the process of transition from film to nucleate boiling. And these are not unstable points because tests for repeatability show that these points can be created either by increasing or decreasing the heat flux. Both the data of Chang and Witte and our experiments verify this.

Whether or not these intermediate temperature points represent a form of transition boiling or true nucleate boiling could not be ascertained from our visual observations, and remains to be investigated.

The Relationship of q_{max} to q_{min}

Figure 12 is a plot of the observed ratio of q_{min} to q_{max} is terms of subcooling for various velocities. The data of Yilmaz and Westwater (1980) at very low subcooling levels are shown for comparison. Yilmaz' data were obtained under uniform surface temperature conditions, with a heater that used condensing steam as the heat source within a copper tube. Even though our data were obtained under constant flux conditions, it seems appropriate to compare the two data sets.

Figure 12 summarizes the effect that we set out to investigate - i.e., whether or not high levels of subcooling and velocity could bring the two transition points closer together. It is clear that subcooling has a dramatic effect in this regard, as our data

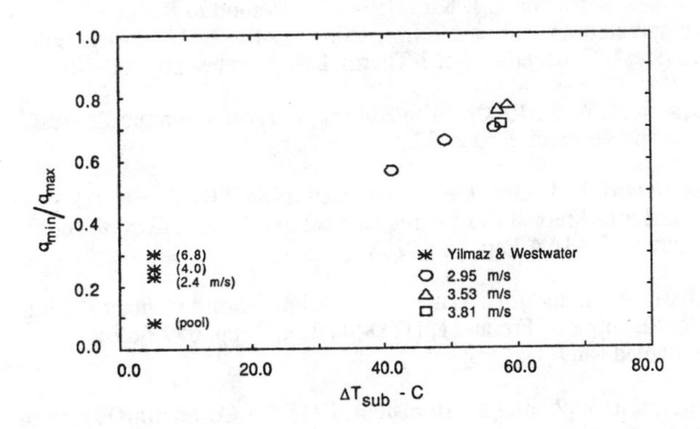

Fig 12. The influence of subcooling on the ratio of minimum to peak heat flux.

clearly show the trend toward unity as subcooling is increased. Although our experiments did not cover a wide range of velocities, it seems clear that subcooling has a much more profound influence on the ratio of q_{min} to q_{max} than does velocity.

CONCLUDING STATEMENTS

Flow boiling of Freon-113 over 0.635-cm cylindrical heaters was studied at subcooling levels up to 58 C and at velocities up to 3.81 m/sec. The experiments show that even at these high levels of subcooling and velocity, a jump-transition occurs as the system goes from nucleate to film boiling and back from film to nucleate boiling. However, the maximum and minimum heat fluxes are brought quite close together by high levels of subcooling. Thus our hope that transition boiling might be eliminated during highly subcooled, high velocity flows continues to hold promise, although the conditions encountered in these experiments were not sufficient to allow such behavior. Our data, taken along with previous measurements, indicates that the ratio of q_{min} to q_{max} is a stronger function of subcooling than velocity.

Insight into the manner in which the transition from nucleate to film boiling and vice versa was also obtained. When the heat flux was increased in slow, precisely controlled increments, film boiling was always observed to occur first in the wake section at the top of the heater, eventually spreading completely over the heater as the flux was increased. The reverse behavior occurred during decreases in heat flux coming from high to lower temperatures. Film and nucleate boiling can coexist on the heater surface and this leads to discontinuities in temperatures at certain locations on the heater because of the proximity of the thermocouple to both regions of film and nucleate boiling. The thermocouple seeks an average temperature between the two extremes of film and nucleate boiling. Thus we have verified the speculation of Chang and Witte that such behavior could occur in a stable manner on electrically heated surfaces.

Repeatable temperature drops in highly subcooled Freon-113 were also measured in the nucleate boiling region. The heat transfer data resemble the "temperature overshoot" measured by many investigators of pool boiling, but no significant increases in bubble nucleation were observed that might explain such temperature drops.

ACKNOWLEDGEMENTS

This work was performed with support from DoE Office of Basic Energy Sciences grant DE-FG05-88ER13893 and NSF grant CBT-8803569.

REFERENCES

Anderson, T.M., and I. Mudawwar, 1988, "Microelectronic Cooling by Enhanced Pool Boiling of a Dielectric Fluorocarbon Liquid", Presented at the 1988 National Heat Transfer Conference, Houston.

Bergles, A.E., and C.J. Kim, 1988, "A Method to Reduce Temperature Overshoots in Immersion Cooling of Microelectronic Devices", Proceedings of I-Therm, Los Angeles, pp. 100-105.

Bradfield, W.S., 1960, "Film Boiling on Hydrodynamic Bodies", Convair Research Report 37.

Broussard, R.A., and J.W. Westwater, 1984, "Boiling Heat Transfer of Freon-113 Flowing Normal to a Tube: Effect of Tube Diameter", AIAA Paper 84-1708.

Chang, K.H, and L.C. Witte, 1988, "Liquid-Solid Contact During Flow Boiling of Freon-11, HTD-96, Vol. 2, pp. 659-665. Accepted for J. Ht. Trans.

Churchill, S.W., and M. Bernstein, 1977, "A Correlating Equation for Forced Convection from Gases and Liquids to a Circular Cylinder in Crossflow", J. Ht. Trans., Vol. 99, pp. 300-306.

Jung, C., and A.E. Bergles, 1989, "Evaluation of Commercial Enhanced Tubes in Pool Boiling", Report DOE/ID/12772-1, Rensselaer Polytechnic Institute.

Lienhard, J.H., and R. Eichhorn, 1976, "Peak Boiling on Cylinders in a Cross Flow", Int. J. Ht. Mass Trans., Vol. 19, pp. 1135-42.

Stevens, J.W., and L.C. Witte, 1973, "Destabilization of Vapor Film Boiling Around Spheres", Int. J. Ht. Mass Trans., Vol. 16, pp. 669-678.

Ungar, E.K., and R. Eichhorn, 1988, "A New Hydrodynamic Prediction of Peak Heat Flux from Horizontal Cylinders in Low Speed Upflow", ASME HTD-96, Vol. 2, pp. 643-657.

Witte, L.C., 1967,"Heat Transfer from a Sphere to Liquid Sodium During Forced Convection", Argonne National Lab Report ANL-7296. See also, L.C. Witte, "An Experimental Study of Forced Convection Heat Transfer from a Sphere to Liquid Sodium", J. Ht. Trans., Vol. 90, pp. 9-12.

Yilmaz, Y., and J.W. Westwater, 1980, "Effect of Velocity on Heat Transfer to Boiling Freon-113", J. Ht. Trans., Vol. 102, pp. 26-31.

AN EXPERIMENTAL INVESTIGATION INTO THE INFLUENCE OF HEATING PLATE THICKNESS ON THE HEAT TRANSFER RATE IN FLOW BOILING

M. A. R. Akhanda
Mechanical Engineering Department
Bangladesh University
Dhaka, Bangladesh

D. D. James
Mechanical Engineering Department
UMIST
Manchester, United Kingdom

ABSTRACT

An experimental investigation of flow boiling was conducted to study the effect of heating plate thickness on the boiling heat transfer rate. Wide ranges of flow velocities and inlet subcoolings were employed together with a heating surface thickness variation from 0.54mm to 1.60mm. The effect of surface micro-roughness were isolated by maintaining a constant surface roughness for all specimens employed in this study. The heating surfaces were also 'aged' prior to all tests. It was found that, at fully developed boiling, the heat transfer rate for a given superheat temperature changed only marginally with heating plate thickness over the range investigated.

NOMENCLATURE

a	channel aspect ratio	
A	area	m^2
C	specific heat	kJ/kgK
k	thermal conductivity	kW/mK
q	heat transfer rate	kW
Ra	centre line surface roughness average (CLA)	μm
t	thickness of the heating plate	mm
T	temperature	K
V	flow velocity	m/s

Greek Symbols

α	thermal diffusivity, $k/\rho C$	m^2/s
ΔT_{sat}	wall superheat temperature, $T_w - T_{sat}$	K
ΔT_{sub}	inlet subcooled temperature, $T_w - T_b$	K
ρ	density	kg/m

Subscripts

b	bulk
sat	saturation
sub	subcooled
w	wall

INTRODUCTION

It is well established that the heat transfer rate in nucleate boiling is dependent on many factors. Surface micro-roughness, surface 'ageing, thermal properties of the heat transfer surface and heat transfer surface thickness must all be considered as well as the fluid dynamic aspects of the problem. The existing state of knowledge is such that there is still uncertainty concerning the effect of heating surface thickness on the heat transfer coefficient in nucleate boiling. A number of workers have investigated this effect in pool boiling and among these are Sharp (1964) who carried out experiments using heater walls of various materials. He concluded that the rate of heat transfer during boiling could be directly related to the thermal conductivity of the heating wall divided by the square root of its thermal diffusivity ($k/\sqrt{\alpha}$). He observed that, as this ratio decreased, so did the boiling heat transfer coefficient. He also hypothesized that the thickness of a heater wall influenced the heat transfer rate during boiling. Further, he proposed that, as the thickness of the heater wall decreased, the boiling heat transfer coefficient should also decrease. The same result, based upon the mathematical model of the micro-layer evaporation theory, was predicted by Dzakamic and Frost (1970). Magrini and Nannei (1975) investigated, under pool boiling conditions, the influence of the thickness of an internally heated surface on the heat transfer coeffient in pool boiling. These workers found that the heat transfer coefficient increased as the heater thickness decreased. This effect was observed only in heaters of a thickness below a certain value and this value (of thickness) varied when different materials were used for the heating plate. Grigoriev et al (1975) conducted pool boiling experiments at atmospheric pressure with cryogenic fluids. Different surface materials of various thicknesses in the range of 0.20 to 20 mm were tested. They found that, at a given heat flux, the wall superheat for a thin-walled surface was higher than that for a thicker-walled surface and the difference increased with an increase in the heat flux. Chuck and Myers (1978) carried out pool boiling experiments to determine the effect of heater plate thickness on boiling heat transfer coefficients for water, ethanol and n-heptane. Stainless steel plates with thicknesses of 0.025 mm, 0.051 mm and 0.13 mm were employed. The heat flux during boiling ranged from 30,000 to 100,000 W/m². It was found that the heat transfer coefficient did increase somewhat with increasing plate thickness for the larger values of ΔT_{sat} (15°C above) and for smaller values the trend was reversed. Del Valle (1980) investigated the effect of wall thickness on the heat transfer rate in flow boiling of water at atmospheric pressure. The test surfaces were stainless steel strips with thicknesses of 0.08 mm, 0.13 mm and 0.20 mm. He found that, at a given heat flux, the corresponding wall superheat became higher as the wall thickness decreased. The effect of wall thickness on CHF was also investigated and it was found that an increase in CHF was observed between the 0.08 mm and the 0.20 mm thick wall ranging from 38% to 57%. Del Valle also developed an empirical expression for CHF which included wall thickness as a parameter and the expression indicated a limiting value for wall thickness affecting the CHF (~ 0.50 mm). He reported that the surface micro-roughness of the heating plate was given by Ra = 0.139 ±20% μm.

From the literature, it is quite clear that most of the work done in this field has been carried out under pool boiling conditions except for the work of Del Valle (1980). All investigations except for Grigoriev et al (1975) were restricted to specimens having thicknesses of less than 0.20 mm. Although Del Valle (1980) suggested that there is no effect of heating plate thickness after 0.50 mm, there is a lack of experimental evidence in flow boiling to support this view. With the exception of Del Valle, no mention has been made of the additional effects of surface roughness.

DESCRIPTION OF THE EQUIPMENT

Fig. 1 presents schematically the experimental apparatus and measuring equipment. The flow circuit comprised a closed loop through which the working liquid was circulated. The rig incorporated a test section, storage and degassing tank, a circulating pump, and other auxilliary units.

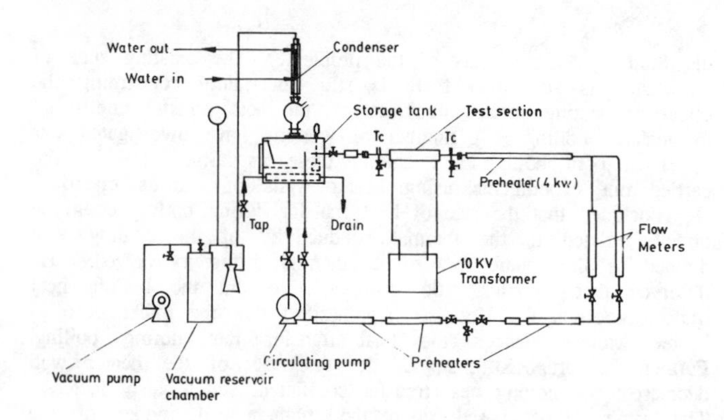

Fig. 1 Schematic Diagram of the Experimental Apparatus

Heating of the test specimens was effected by the passage through them of a high ac current (0-700 amps) at a low voltage (0-6 volts) from a heavy duty 10 kVA transformer which was regulated by a double variac. To obtain the total heat input to the heating surface, current was measured by a precision ammeter through a current transformer and voltage across the heating surface was measured with an ac digital voltmeter. The maximum error involved in the measurement of heat flux was approximately 4%. Surface temperature was measured using seventeen 36 SWG Chromel-Alumel thermocouples spot welded to the back of the heating plate and was recorded by a Data logger with teletype. All thermocouples were calibrated to within ± 0.16&. A correction involving the application of the steady state one-dimensional heat conduction equation with internal heat generation was made to evaluate the surface temperature. The inlet and outlet bulk temperatures were recorded by two thermocouple probes. The flow rate was controlled by adjusting, as required, the gate valves (throttle valves), the diaphragm valves (meter valves) and the by-pass valve installed in the circuit. It was measured by two rotameters in parallel. The system pressure was maintained constant by a needle valve connected to the vacuum reservoir chamber to which a vacuum pump was connected and was monitored by a pressure gauge and a calibrated pressure transducer. The pressure difference between the inlet and outlet of the test section (which was very small) was measured by a water column manometer. Inlet bulk temperature was maintained constant by regulating the cooling water flow rate through the glass cooler in the tank (when the inlet bulk temperature was higher than the required value) and by regulating the preheater installed upstream of the test section by a double variac (when this temperature was lower).

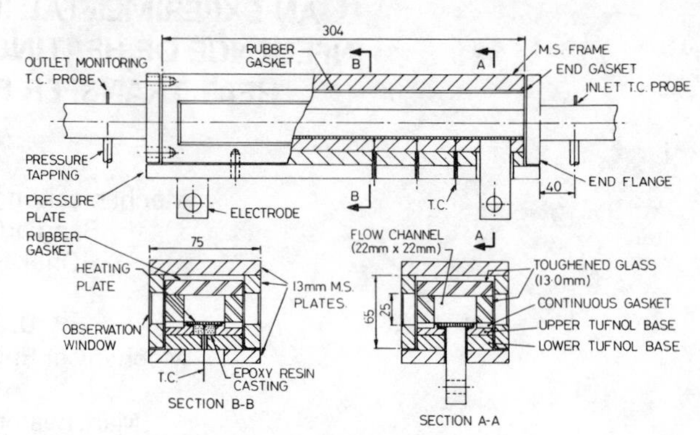

Fig. 2 Test Section

Details of the test section are shown in Fig 2. It was 304 mm in length with a cross section of 22 mm x 22 mm. Two pressure taps were provided, each of which was installed 40 mm away from the test section and these were connected to a pressure gauge and a pressure transducer.

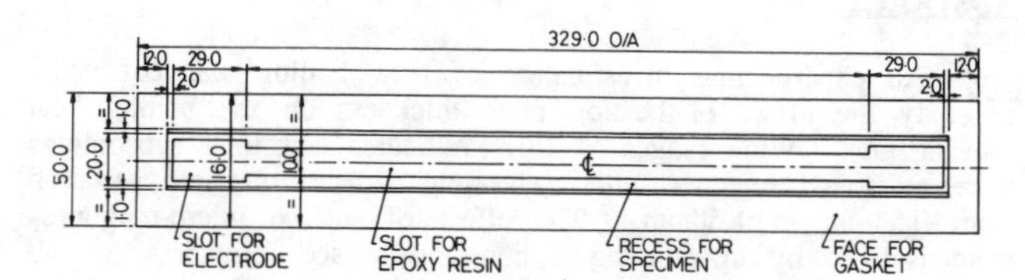

Fig. 3(a) Details of Upper Tufnol Base

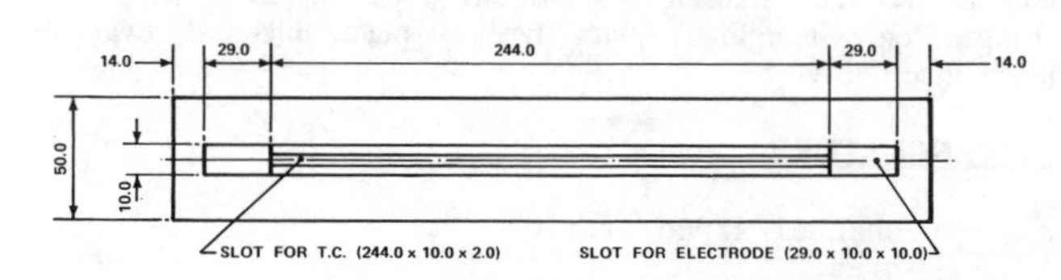

Fig. 3(b) Details of Lower Tufnol Base

The upper and lower bases were manufactured from 'Tufnol' (Phenolic resin, ASP brand) material. These are shown in detail in Figs 3(a) and (b). The heating plate was located in the recess of the upper base and held in position by means of heat resistant Araldite 2004. The thermal shock resistant Epoxy Resin (Araldite MY 750, hardener HT 972 and a crack resistant filler-mica flour) was then cast at the back of the heating surface through the specified slot for reinforcement. Before carrying out the above two operations, thermocouples were spot welded to the back of the heating plate. These two bases were used as backing and insulating materials.

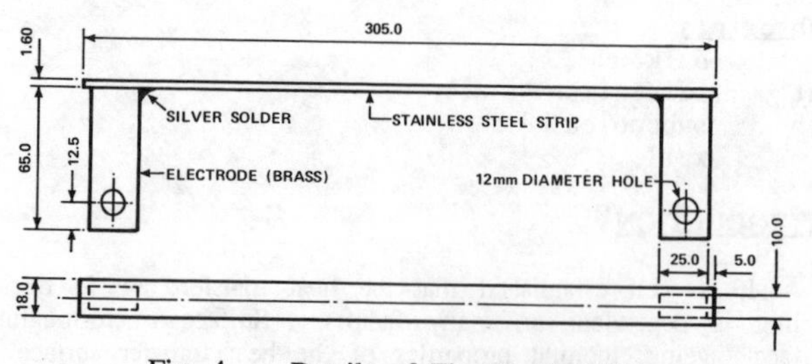

Fig. 4 Details of Flat Heating Plate

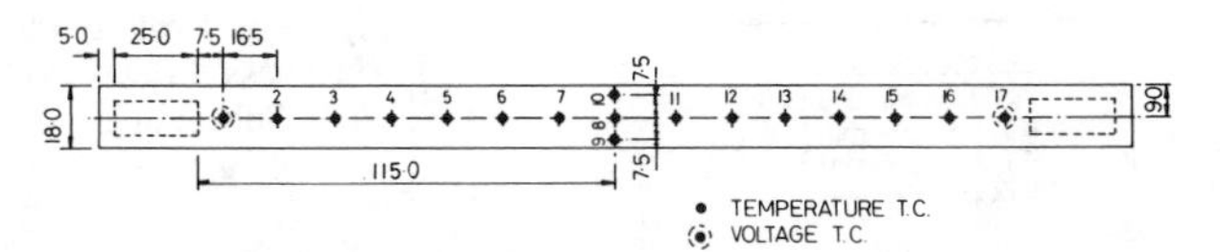

Fig. 5 Position of Thermocouples and Voltage Tappings

The heating surface arrangement is shown in Figs 4 and 5. Brass electrodes of 65 mm x 25 mm x 10 mm were silver soldered 5 mm away from both ends of each test specimen. The dimensions of the heating plates employed in this work are given in Fig 6 and also in the following table:

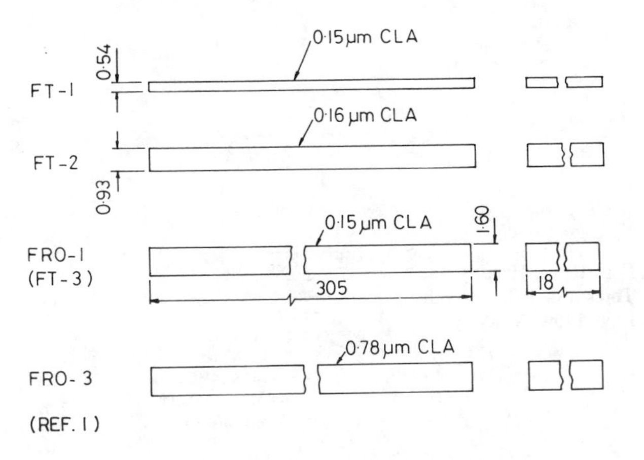

Fig. 6 Dimensions of Test Specimens

Table 1

Surface	CLA Surface Roughness Ra μm	Thickness t mm
FT - 1	0.15 ± .02	0.54
FT - 2	0.15 ± .02	0.93
FT - 3	0.15 ± .02	1.60
'FRO - 3' (ref 1)	0.78 ± .05	1.60

'FT' - Flat smooth surface changing thickness
'FRO' - Flat roughened surface

All the heating plates (FT-1, 2 and 3) were manufactured from stainless steel type EN58E (AISI 304) of thickness 24 SWG, 20 SWG and 16 SWG respectively, These specimens were polished with a series of wet and dry emery paper of grades 320, 400, 500 and 600 respectively. The surfaces were then polished with a liquid metal polish until they reached a mirror finish. A more detailed description of the experimental apparatus is given in Akhanda & James (1985).

RESULTS AND DISCUSSION

The experimental programme to investigate the effect of the thickness of the heating plate on the boiling heat transfer rate was carried out under the conditions shown in Table 2.

Table 2

System Pressure	Atmospheric
Working liquid	Distilled & deionized water
Test section geometry	A channel of rectangular section with one side heated only
Channel aspect ratio	a = 1
Heat transfer surfaces	FT-1 (t = 0.54mm), Ra=0.15μm FT-2 (t - 0.93mm), Ra=0.15μm FT-3 (t = 1.60mm), Ra=0.15μm
Material	Stainless Steel EN58E (AISI 304)
Inlet Subcoolings	5°C & 30°C
Flow velocities	0.20 m/s, 0.35 m/s, 0.70 m/s 1.10 m/s & 1.40 m/s
Heat flux	0.0 - 500 kW/m²

Heat transfer results of the experiments with 'FT' surfaces in terms of heat flux, superheat temperature, flow velocity and inlet subcooling are shown from Fig 7 to Fig 11.

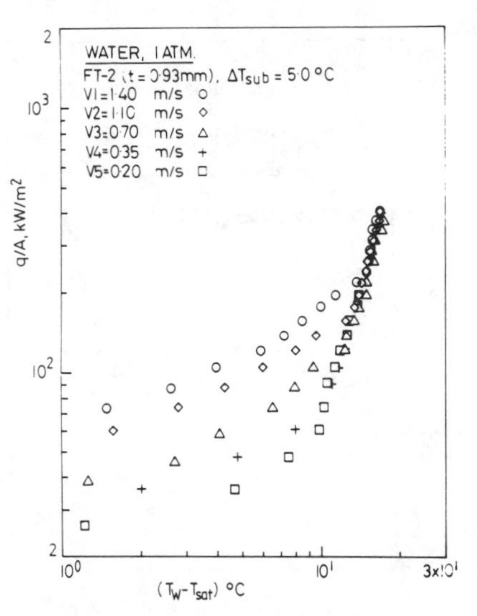

Fig. 7 Effect of Flow Velocity

Fig 7 is a typical plot of heat flux (q/A) versus wall superheat (ΔT_{sat}) for different flow velocities.

Fig 8 shows the effects of inlet subcooling. These effects of velocity and inlet subcoolings are in good agreement with the results of Del Valle (1980) and Akhanda & James (1984). Pressure drop across the short test section was also measured but no significant changes were observed over the experimental range investigated.

The effect of the thickness of the heating plate on the heat transfer rate in fully developed sub-cooled flow boiling was observed for water. Data were obtained from three flat 'FT'

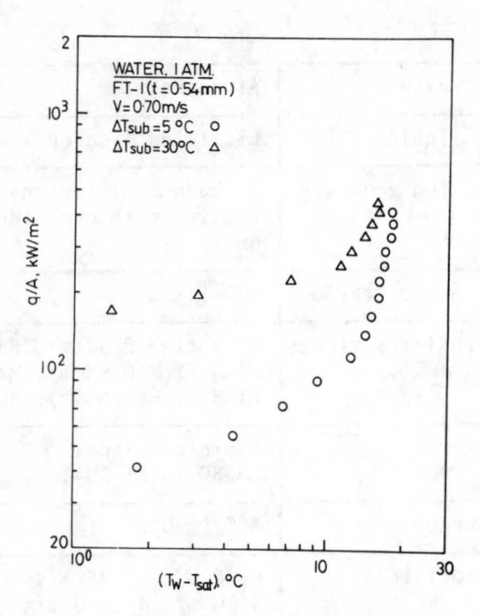

Fig. 8 Effect of Inlet Subcooling

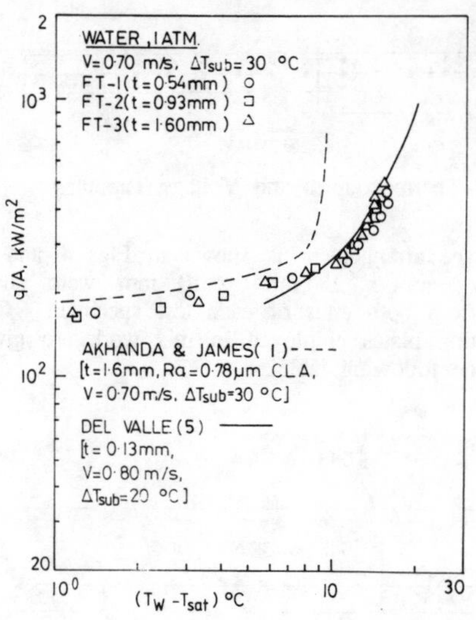

Fig. 9(b) Heating Plate Thickness Effects with High Subcooling

surfaces in which the effects of surface roughness were isolated by maintaining constant surface micro-roughness at 0.15 μm ± 0.02μm. Test results in terms of heat flux (q/A) and wall superheat (ΔT_{sat}) for three specimens at different inlet subcoolings under atmospheric pressure conditions and at a typical constant flow velocity of 0.70 m/s are given in Figs 9(a) and (b).

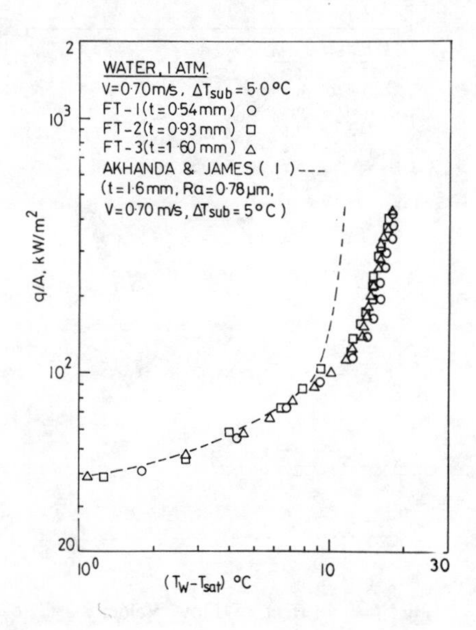

Fig. 9(a) Heating Plate Thickness Effects with Low Subcooling

In each figure, the data for subcooled flow boiling with the optimum roughened surface 'FRO-3' from Akhanda & James (1984) under identical experimental conditions are included for comparison. It should be noted that the data presented in the above figures were recorded after the surfaces had been 'aged' sufficiently as mentioned in Akhanda & James (1985). It will be observed that for all inlet subcoolings under fully developed boiling conditions, the heat transfer rate for a given superheat temperature changed only marginally for a heating plate thickness greater than 0.54 mm, thus confirming the prediction of Del Valle (1980) over a reasonably wide range of heating plate thicknesses.

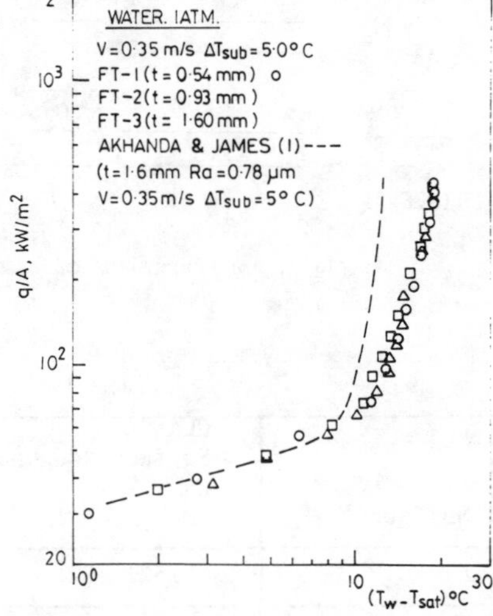

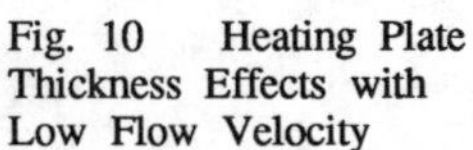

Fig. 10 Heating Plate Thickness Effects with Low Flow Velocity

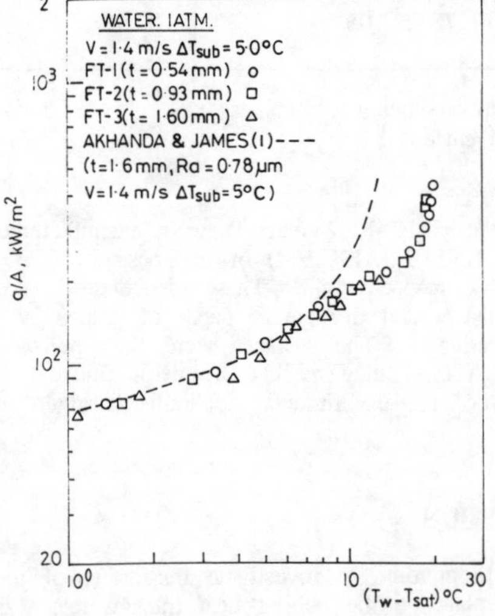

Fig. 11 Heating Plate Thickness Effects at High Velocity

Figs 10 and 11 are the typical representations of the data recorded for the same surfaces with an inlet subcooling of 5°C and flow velocities of 0.35 m/s and 1.40 m/s respectively. It will be observed that the trends exhibited are identical with those shown at a velocity of 0.70 m/s.

The values of 'n' in the equation $q/A \sim \Delta T n_{sat}$ for 'FT' surfaces were found to be constant at 4.10. This effect was expected because heat transfer performance did not vary much among these specimens. The values of 'n' lay almost within the range of Bruzzi (1969) which would be expected since the surface micro-roughness of the specimens employed here were almost the same as those of Bruzzi (1969).

CONCLUSIONS

Experiments were carried out in forced convective boiling for a wide range of flow velocities and inlet subcoolings to investigate the effect of heating plate thickness on the boiling heat transfer rate. Over the range investigated, it was observed that, under fully developed boiling conditions, the heat transfer rate for a given superheat temperature changed only marginally. This result is in keeping with the expectations of existing semi-theoretical predictions.

ACKNOWLEDGEMENTS

This work was carried out in the Department of Mechanical Engineering, UMIST, Manchester, UK. One of the authors, M.A. R. Akhanda is grateful to the Commonwealth Scholarship Commission, UK for the award of a Commonwealth Scholarship which enabled him to carry out this work.

REFERENCES

Akhanda, M.A.R. and James, D.D., (1984), "Forced Convection Boiling from Roughened Surfaces", *11th Annual Research Meeting on Heat Transfer and Catalysis and Catalytic Reaction,* I Chem E, Bath, UK.

Akhanda, M.A.R. and James, D.D., (1985), "Surface Aging and Reproducibility of Data in Forced Flow Boiling" *8th National Heat & Mass Transfer Conference,* Indian Society of Heat and Mass Transfer, Visakhapatnam, India.

Bruzzi, S., (1969), "Forced convection boiling heat transfer in a rectangular duct heated on one side", PhD Thesis, Imperial College, London.

Chuck, T.L. and Myers, J.E., (1978), "The Effect of Heater Plate Thickness on Boiling Heat Transfer Coefficients", *Int J Heat Mass Transfer,* Vol 21, pp 187-191.

Del Valle, V.H.M., (1980), "Flow Boiling near the Critical Heat Flux", D Phil Thesis, University of Oxford.

Dzakamic and Frost, (1970), "Vapour Bubble Growth in Saturated Pool Boiling Microlayer Evaporation of Liquid at the Heated Surface", *Proceedings, 4th International Heat Transfer Conference,* Paris-Versailles, Paper No B 2.2.

Grigoriev, V.A. et al (1975), *Trudy MEI* (Trudy-Moskovskoyo Energetischesecgo Instituta), Series 268, p.53.

Magrini, U. and Nannei, (1975), "On the Influence of the Thickness and Thermal Properties of Heating Walls on the Heat Transfer Coefficients in Nucleate Pool Boiling", Trans ASME, *J Heat Transfer,* Vol 97C, p173.

Sharp, R.R., (1964). "The Nature of Liquid Film Evaporation during Nucleate Boiling", *NASA Report,* TNO-1997.

RIB ORIENTATION EFFECTS ON HEAT TRANSFER PERFORMANCE IN FORCED CONVECTION BOILING

M. A. R. Akhanda
Department of Mechanical Engineering
Bangladesh University
Dhaka, Bangladesh

D. D. James
Department of Mechanical Engineering
UMIST
Manchester, United Kingdom

ABSTRACT

Experimental studies of the enhancement of heat transfer in convective boiling from rectangular ribbed surfaces is presented. Particular attention is focussed on the effects of rib orientation to the flow direction for a range of flow velocities and inlet subcoolings. Rib orientation to the direction of flow varied from 0° to 135° and rib height, length and spacing were maintained constant at 0.50 mm, 1.0 mm and 0.50 mm respectively. The effects of surface roughness were isolated by providing a constant surface micro-roughness for all specimens, and the effects of 'ageing' of the heat transfer surface were established prior to the commencement of all tests. The experimental data yielded by the investigation are compared with those for a smooth and a roughened flat surface and it is observed that the optimum heat transfer was achieved with transverse ribs, i.e. those oriented at 90° to the direction of the flow.

NOMENCLATURE

a	channel aspect ratio	
A	flat projected area of the heating surface	m^2
C_p	specific heat at constant pressure	kJ/kgK
e	rib height	mm
g	acceleration due to gravity	m/s^2
h	heat transfer coefficient, $(q/A)/(T_w-T_{sat})$	kW/m^2K
L	rib length	mm
Nu	Nusselt number	
q	heat transfer rate	kW
Ra	centre line surface roughness average (CLA)	m
s	spacing between ribs	mm
T	temperature	K
ΔT_{sub}	inlet subcooling, $(T_{sat} - T_b)$	K
ΔT_{sat}	wall superheat, $(T_w - T_{sat})$	K
V	flow velocity	m/s

Greek Symbols

λ	latent heat of vaporisation	kW/mK
ρ	density	kg/m
θ	angle with the direction of flow	degrees

Subscripts

calc	calculated
expt	experimentally obtained
ℓ	liquid condition
sat	saturation
sub	subcooled condition
v	vapour
w	wall

INTRODUCTION

Much attention has been paid recently to the study of enhanced heat transfer employing surface geometry variations in the form of fins and spines. Among the possible benefits resulting from such a study are the saving of material and the production of more compact heat exchangers especially where rectangular flow geometries are concerned. Significant advances have been made by investigating the use of surface geometry variations to promote high performance heat transfer characteristics in pool boiling but, there is little information available in the literature on the heat transfer performance for flow geometries of circular cross section, and even less for rectangular flow geometries.

Heat transfer augmentation employing finned surfaces has been studied by many investigators including Haley and Westwater (1965a), Haley (1965b), Klein and Westwater (1971), Simon-Tov (1970), Bondurant and Westwater (1971), Lai and Hsu (1967) and Hsu (1968). The influence of smaller surface protuberances (ribbed surfaces) of the order of a bubble diameter at departure at atmospheric pressure would be expected to exhibit different characteristics in boiling from those associated with the larger fins and spines referred to above. Dunn and Reay (1976) reported heat transfer data from a variety of grooved heat pipe walls and grooved wick structures of an arterial type heat pipe. Gorenflo (1966) carried out experiments on nucleate boiling from copper tubes with small circumferential fins using refrigerant R-11 at pressures above atmospheric. Fin height was varied from 1.5 mm to 3.5 mm, fin spacing from 1 to 3 mm and fin thickness from 0.3 to 0.8 mm, whilst the surface roughness of the tube varied between 0.06 and 0.55 µm and the ratio of the total surface area to the projected area varied from 1 to 4.92. He compared the performance of these finned tubes with that of a smooth one using both total and projected areas and found an improvement in heat transfer with finned tubes. He also suggested that, for the finned tubes, a critical spacing of approximately twice the bubble departure diameter existed. Tubes with fin spacing above and below this value exhibited improved heat transfer, whilst a diminution in heat transfer was observed for tubes with this critical spacing. Gorenflo (1968) also considered the behaviour of other refrigerants in the same tubes over a wider range of pressure (0.1 bar to 9.0 bars). In both studies (Gorenflo (1966,1968)) the heat transfer coefficient was related to the heat flux and a pressure function as $h \sim (q/A)^n f(p)$

A relation for this pressure function suggested by Danilova (1965) was extended to include surface geometry variations as follows:

$$f(p, A_t/A_p) = 0.14 + 2.2\,(p/p_{cr}) \quad \frac{A_t}{A_p}$$

where A_t and A_p are the total and projected areas respectively. Hesse (1973) compared boiling of refrigerant R-114 from a grooved nickel tube with boiling from a smooth tube having a surface roughness of 0.61 μm at pressures of 3 bars and 6 bars. The grooves were 0.5 mm deep, 0.6 mm apart and 0.40 mm thick, having a surface roughness average of 0.42 μm. The heat transfer from the grooved tube was found to be better than the smooth tube, with both total and projected area considerations at a pressure of 3 bars. Heat transfer from the grooved tube was improved when the projected area was considered at a pressure of 6 bars.

Joudi (1977) investigated experimentally nucleate pool boiling from stainless steel surfaces in water, refrigerant R-113 and methanol. He used a number of ribbed surfaces, all having the same surface micro-roughness. Generally, surface ribs of triangular cross section exhibited better heat transfer performance in boiling than surfaces with rectangular cross-section. Also, surfaces with a wide clearance between ribs which facilitated easier liquid and vapour flow showed an improved boiling performance compared with specimens with narrow rib spacing. Webb (1981) surveyed the evolution of special surface geometries to promote high performance nucleate boiling. Akhanda and James (1984a) investigated experimentally the forced convection boiling of water from one side of a rectangular duct having a channel aspect ratio of 1. Boiling occurred from transverse rectangular ribbed surfaces all having the same rib height of 0.50 mm and length of 1.0 mm, while spacing between ribs ranged from 0.50-3.50 mm. The results were compared with a smooth surface and an optimum heat transfer performance was achieved with rib spacing of 1.0 mm. Akhanda and James (1985a) also report on the relative effect of transverse and longitudinal ribbing of the heat transfer surface in subcooled flow boiling for an identical channel geometry. These findings showed that an optimum transverse ribbed surface yielded a better heat transfer performance than an optimum longitudinal ribbed surface and that both showed an enhanced heat transfer rate compared with an optimum roughened surface without ribs as given in Akhanda and James (1984b). A review of the development in enhanced heat transfer is presented by Bergles et al (1981). The author also summarised the different techniques developed for the augmentation of heat transfer in boiling.

The work described below comprises a study of heat transfer augmentation using surface geometry variation of the order of a bubble diameter at departure under atmospheric conditions. A rectangular flow geometry is employed with heat transfer occurring from one side of the duct to an internally flowing fluid.

GENERAL DESCRIPTION OF THE EQUIPMENT

The experimental apparatus and measuring equipment are shown schematically in Fig 1. The rig consisted of a closed loop system incorporating a test section, a storage and degassing tank, circulating pump, condenser, vacuum pump, vacuum reservoir chamber, flow meters and other auxiliaries. The system pressure was regulated by a needle valve connected to the condenser. Four

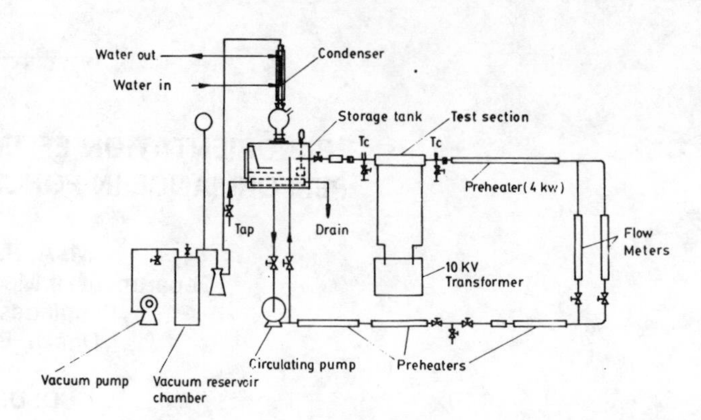

Fig. 1 Schematic Diagram of the Experimental Apparatus

preheaters were used to control the temperature of the working fluid and control of the flow rate was achieved by operating a number of valves upstream of the flow meters. All components in contact with the working liquid were manufactured from materials which could adequately withstand the liquid boiling temperature and also resist corrosion.

Test Section

Figure 2 shows the test section in detail. Two pressure tappings were provided on the bottom wall of the inlet and outlet ducts. Each tapping was installed 40 mm away from the test section and was connected to a pressure gauge and a calibrated pressure transducer. Pressure drops across the short test section were measured but no significant changes were observed over the experimental range investigated.

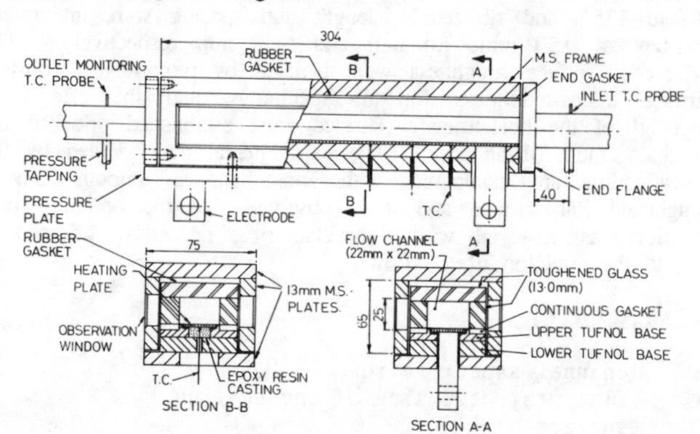

Fig. 2 Test Section

The base plate assembly comprised two components: (a) an upper tufnol base and (b) a lower tufnol base. The heater was attached to the upper 'tufnol' (Phenolic resin - Brand ASP) base by means of heat resistant Araldite 2004. A slot was machined into the upper tufnol base to receive the heating surface (specimen) so that after assembly, this was flush with the base. The groove beneath the specimen was filled with a casting of epoxy resin (Araldite MY750, hardener HT972 and a crack resisting filler-mica flour) to strengthen the assembly. The upper tufnol base also contained a recess for a continuous gasket onto which the 'U' shaped glass channel (made up of 13 mm thick toughened glass) and the two end flanges were located. The lower tufnol base was used to reinforce the arrangement and to provide additional insulation. Details are shown in Fig 3(a) and (b).

The Heating Surfaces

The heating plates were manufactured from one single sheet of 16 SWG stainless steel type EN58E (AISI 304) having a

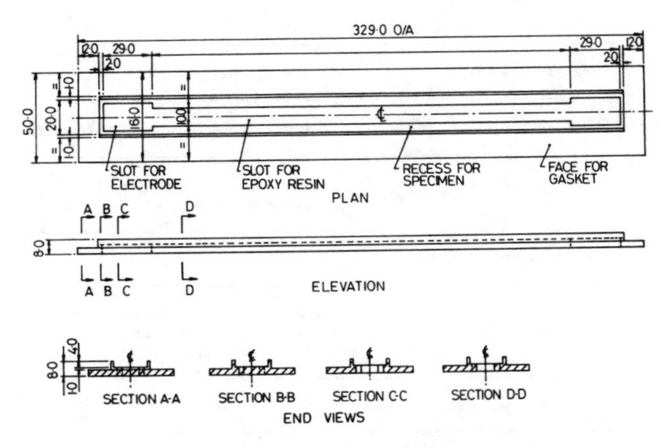

Fig. 3a Details of Upper Tufnol Base

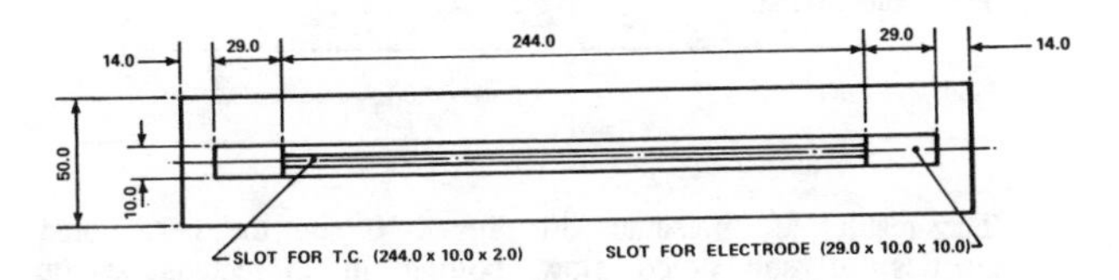

Fig. 3b Details of Lower Tufnol Base

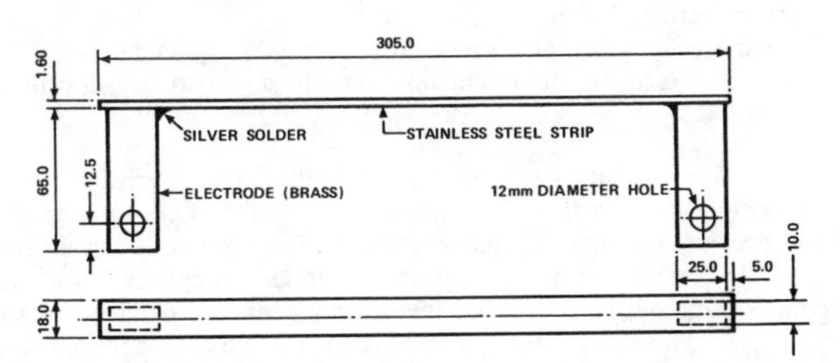

Fig. 3c Flat Heating Plate with Brass Electrodes

thickness of 1.60 mm. After the required surface geometry had been machined onto each specimen, two brass electrodes were silver soldered 5 mm away from the ends of the test specimen as shown in Fig 3(c). The stainless steel surface was finished with vaqua blasting. This consisted of the impingement onto the surface at high pressure of an alumina particle suspension in water. The technique yielded a continuous and reproducible surface micro-roughness of 0.78 μ m CLA in all directions and produced a homogeneous surface texture over the whole surface. Information concerning surface geometry variations is given in Fig 3(d) and dimensions are presented in Table 1.

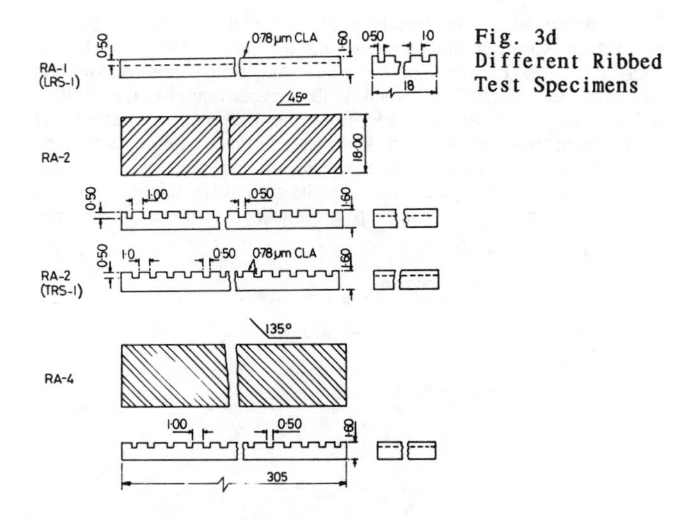

Fig. 3d Different Ribbed Test Specimens

Table 1

Surfaces	CLA Surface Roughness average Ra μm	Rib height 'e' mm	Rib length 'L' mm	Spacing between ribs 's' mm	Rib angles to the direction of flow 'θ' degree
RA-1 (LRS-1) (ref 3)	0.78 ± .05	0.50	1.0	0.50	0
RA-2	0.78 ± .05	0.50	1.0	0.50	45
RA-3 (TRS-1) (ref 1)	0.78 ± .05	0.50	1.0	0.50	90
RA-4	0.78 ± .05	0.50	1.0	0.50	135
FRO-1 (ref 2)	0.15 ± .02	-	-	-	-
FRO-3 (ref 2)	0.78 ± .04	-	-	-	-

RA - Rectangular ribbed surfaces changing 'θ'
FRO - Flat roughened surface
LRS - Longitudinal rectangular ribbed surface changing 's'
TRS - Transverse rectangular ribbed surface changing 's'

Instrumentation and Measurements

Heating of the test surfaces was accomplished by the passage of a heavy ac current (0-1000 amps) at a low voltage (0-6 volts) from a 10 kVA transformer regulated by a double variac. Current was measured by a precision ammeter through a current transformer and voltage was measured directly across the heating surface using two thermocouple wires spot welded near the extreme edges of the stainless steel surface as shown in Fig 3(e). The exact distance between these voltage tappings was determined with a vernier microscope and this dimension was used in the calculation of heat flux. The maximum error involved in the measurement of heat flux was approximately ±4%. The potential drop across this distance was observed on a DVM. Conduction losses to the base assembly were calculated and found to be small (±1%).

The surface temperature was determined from the average indication of 15 Chromel-Alumel thermocouples attached to the base of the test specimen as shown in Fig 3(e). The thermocouples were situated so as to provide a temperature distribution across the length and width (at mid section) of the sample. All were connected to a selector switch and the emfgenerated was measured with a 'solartron' digital voltmeter and also with a data logger with teletype. All thermocouples were calibrated to within ±0.16% A thermocouple junction immersed in melting ice was used as a common cold junction. Figures 3(f) and (g) show actual measurements of temperature distribution along the length and the width (at mid-section) respectively. The data are presented for various inlet subcoolings at a constant velocity.

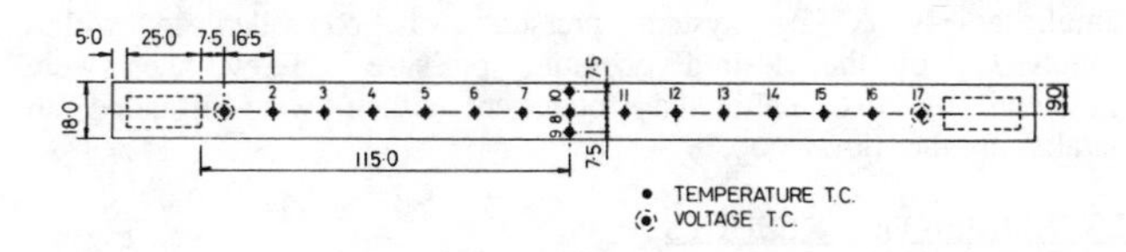

Fig.3e Positions of Thermocouples and Voltage Tappings

The existence of an isothermal region over the central portion of the surface covering the majority of the heat transfer area was observed. The presence of an isothermal temperature distribution over the heating surface enabled the calculation of the boiling surface temperature to be made from the measured temperature at the insulated base of the test specimen, together with a correction using the one-dimensional heat conduction equation with internal heat generation. Since the rib height was relatively small compared with the heating plate thickness (about 1:3), and

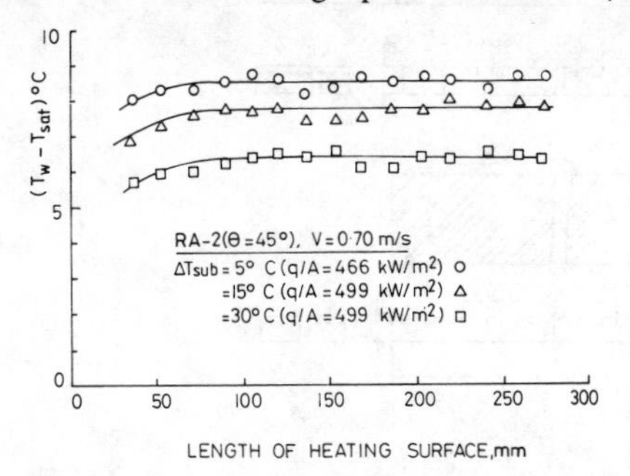

Fig. 3f Temperature Profile along the Length of the Heating Surface

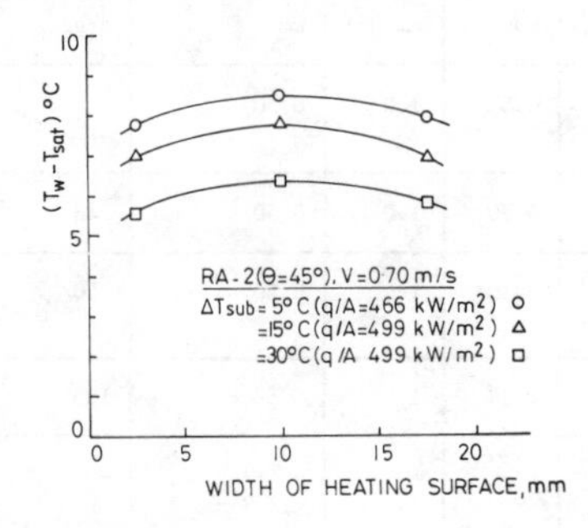

Fig. 3g Temperature Profile along the Width of the Heating Surface

isothermal conditions prevailed at the underside of the heated surface, the error involved in employing the method of the one-dimensional heat conduction equation was considered to be tolerable. Moreover, a numerical analysis using finite difference techniques was carried out to determine the temperature distribution in a two-dimensional mesh with internal heat generation. The results of the analysis indicated that for the worst case, i.e. a heat transfer coefficient of 30 kW/mK and an internal heat generation of 205,000 kW/m^3 , the variation in temperature of the rib base was within acceptable limits of less than 0.5%. The analysis also showed that the temperature varied along the horizontal axis of the specimen only slightly but demonstrated a sharp decrease from the base to the heat transfer surface, ie the temperature varied significantly across the thickness of the specimen but only slightly along its length.

Bulk inlet and outlet temperatures to the test section were measured by thermocouple probes, both thermocouples being of the same type as those used for surface temperature measurement. The system pressure was monitored during experiments by a calibrated pressure gauge and a calibrated pressure transducer simultaneously. The system pressure was controlled to within ±3mm Hg of the desired operating pressure. Flow rates were measured by two calibrated rotameters which were installed in parallel in the flow loop.

EXPERIMENTAL RESULTS

The experimental conditions employed are shown in Table 2.

Table 2

System Pressure	1 Atmosphere
Working Fluid	Distilled & Deionized Water
Test Section	A channel of rectangular cross-section
Channel aspect ratio	a = 1
Test Specimens	Stainless steel [type EN58E (AlSl 304)] - Rectangular ribbed surfaces (RA)
Thickness of the specimen	t = 1.60 mm
Flow velocity	0.20 m/s, 0.35 m/s, 0.70 m/s 1.10 m/s & 1.40 m/s
Inlet subcoolings	5°C, 15°C & 30°C
Heat Fluxes	0.0 - 500 ∿ 600 kW/m^2

The results are presented in Figs 4-10 and the heat transfer characteristics of sub-cooled flow boiling in a rectangular duct heated from one side only are discussed with regard to:

(a) the effects of flow velocity
(b) the effects of inlet subcooling
(c) the effects of rib orientation (ie rib angle) to the direction of flow.

In order to attribute changes in boiling characteristics solely to changes of surface geometry (i.e. ribbing), the strong dependence of the boiling process on surface roughness was first isolated by maintaining a constant surface roughness for all specimens employed. In this investigation as in previous work, Akhanda & James (1984a), (1985a) and (1984b); a constant and reproducible surface finish and roughness was obtained by a vaqua blasting technique with 200 mesh alumina particle size. This particle size was selected for the surface finish of the ribbed surfaces employed since it was observed in Akhanda & James (1984b) that this particle size produced a heat transfer surface of roughness average CLA - 0.78μm, a value which yielded an optimum heat transfer surface. Each of the specimens was 'aged' sufficiently to obtain reproducible data as shown in Akhanda & James (1985a).

Figs 4 and 5 are the typical plots of heat flux (q/A) versus wall superheat (ΔT_{sub}) showing the effects of flow velocity and inlet subcooling respectively. Similar trends were also observed in Akhanda & James (1984a) and (1985a) for ribbed surfaces. In this investigation, the exponent 'n' of the equation: $q/A \sim \Delta T^n_{sub}$ was found to vary from 4.0 to 5.5 for 'RA-surfaces' and the value of 'n' was found to be the highest for the optimum specimen. A similar trend was also observed by Joudi (1977) in his pool boiling experiment with ribbed surfaces. Photographs of the characteristics of bubble behaviour along 'LRS-2' from Akhanda & James (1985a) and 'RA-4' surfaces for an identical velocity and similar heat fluxes are shown in Fig 6(a) and (b).

The effect of rib orientation i.e. rib angle (θ) to the direction of flow are shown in Figs 7(a) to (c). These represent the experimental data yielded by specimens 'RA-1', 'RA-2', 'RA-3' and 'RA-4" at inlet subcoolings of 5 C, 15 C and 30 C respectively and at a constant flow velocity of 0.70 m/s. In order to attribute the heat transfer characteristics solely to changes of rib orientation (θ) to the direction of flow, rib height, rib length and rib spacing were maintained constant. Curves of the data obtained

Fig. 4(a)

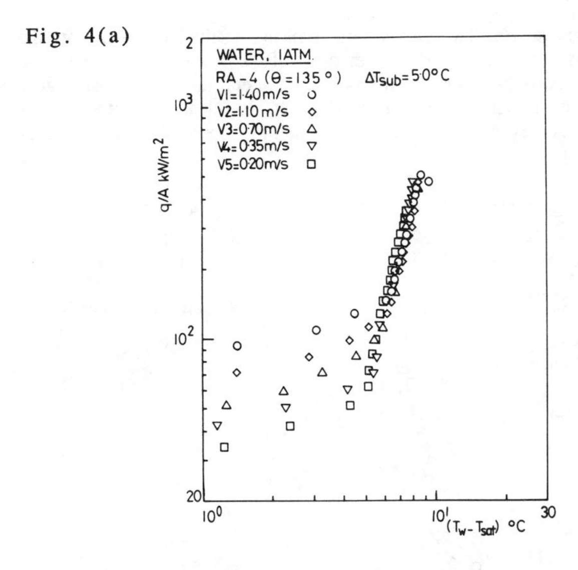

Fig. 4(b)

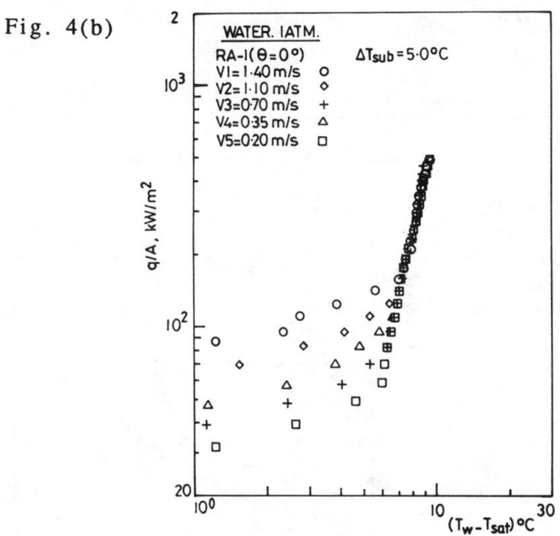

Figs. 4(a) & 4(b) Effects of Flow Velocities

Fig. 5
Effects of Inlet Subcoolings

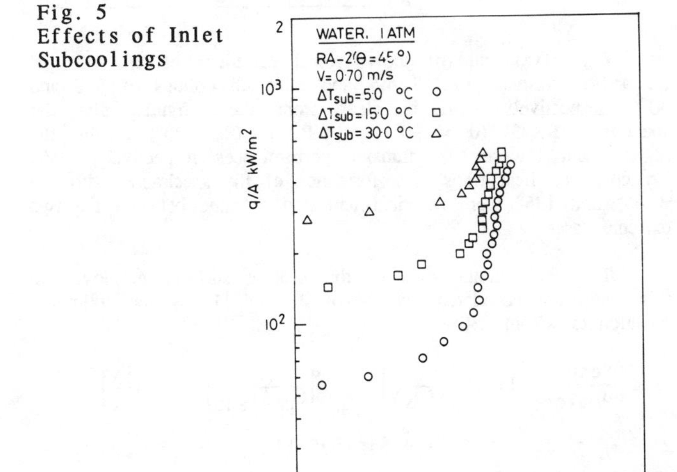

from a polished flat surface (FRO-1) and an optimum roughened surface (FRO-3) from Akhanda & James (1984b) are included in each of the above figures for the purpose of evaluating improvements in heat transfer performance. It is observed from these figures that in fully developed boiling at a constant superheat (ΔT_{sat}), the heat flux (q/A) was highest with specimen 'RA-3', having $\theta = 90°$. As expected, the heat flux was found to be the same for specimens having $\theta = 45°$ and 135°. The minimum heat transfer performance occurred for specimens having $\theta = 0°$, ie a longitudinal rib. It will be observed that in fully developed boiling, the heat flux for a given superheat increased with increasing θ upto $\theta = 90°$ and then decreased. It is clear that the specimen 'RA-3', having $\theta = 90°$, gave the optimum heat transfer performance for the range of 'RA-surfaces' investigated.

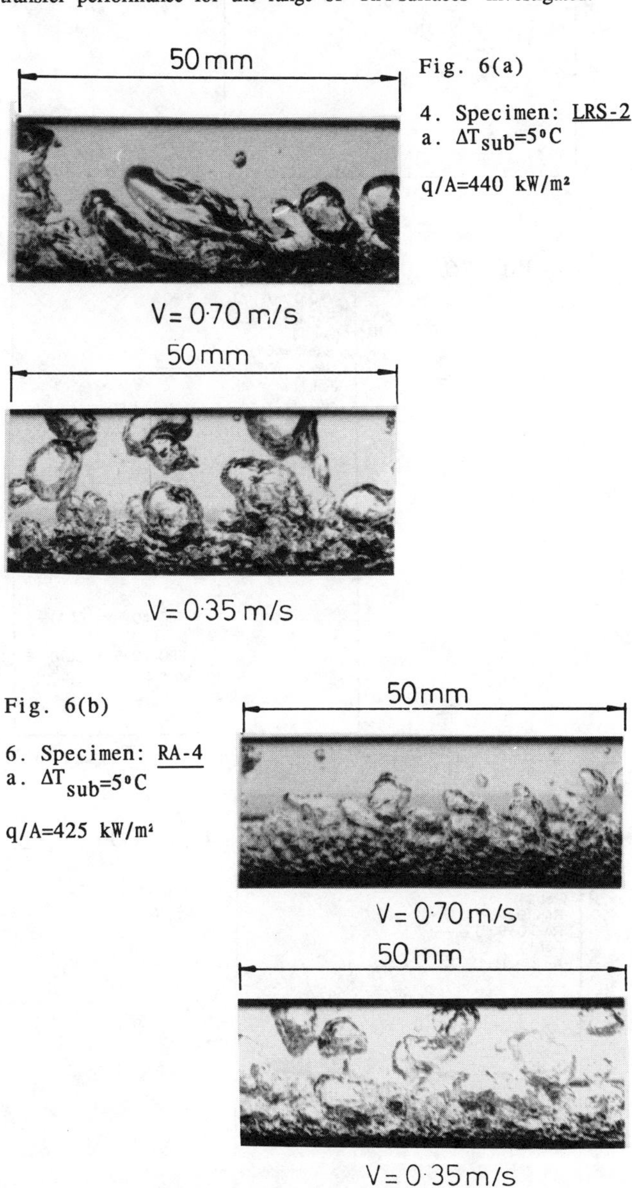

Fig. 6(a)

4. Specimen: LRS-2
a. ΔT_{sub}=5°C

q/A=440 kW/m²

Fig. 6(b)

6. Specimen: RA-4
a. ΔT_{sub}=5°C

q/A=425 kW/m²

Figs. 6(a) & 6(b) Photographic Studies of Bubble Behaviour along the Typical Ribbed Surface

The maximum improvement of heat transfer performance compared with a very smooth surface 'FRO-1' was about 60% for inlet sub-cooling of 5°C, 57% at an inlet subcooling of 15°C and 65% at an inlet subcooling of 30°C all for a heat flux of 500 kW/m and at a constant velocity of 0.70 m/s.

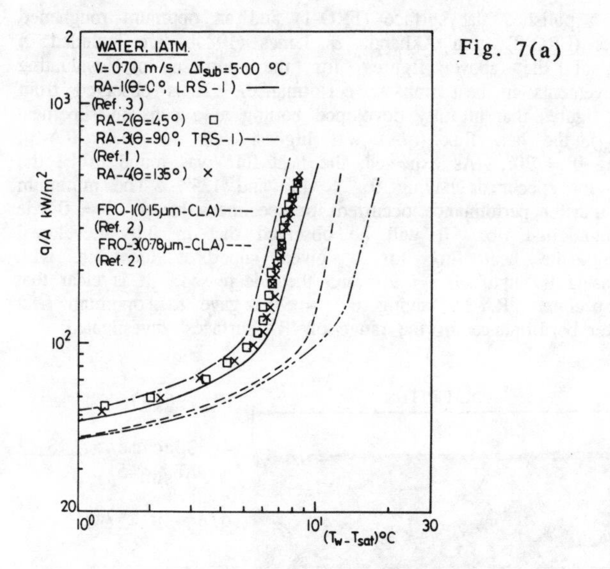

Fig. 7(a)

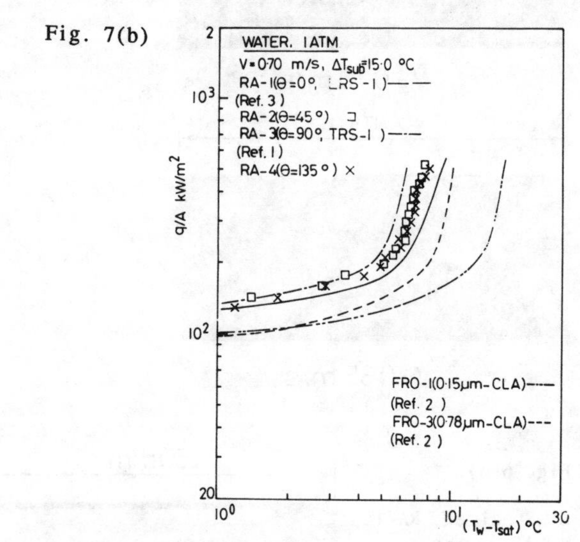

Fig. 7(b)

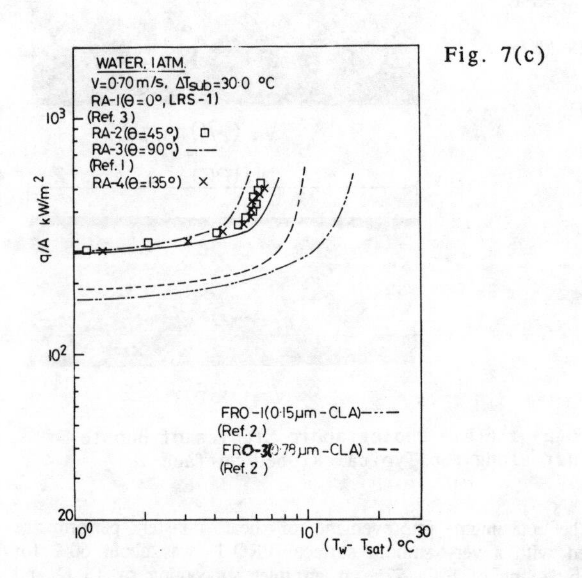

Fig. 7(c)

Figs 7(a)(b) & (c) Effect of Rib Inclination to the Direction of Flow at a Flow Velocity of 0.7m/s

Figures 7(a), 8 and 9 are typical results recorded for the same surfaces with a flow velocity of 0.35 m/s, 0.70 m/s and 1.40 m/s corresponding to Reynolds numbers up to 10^5 and at a constant inlet subcooling of 5° C. It is clear that the trends exhibited at higher velocity are similar to those observed in Figs 7(a) to (c).

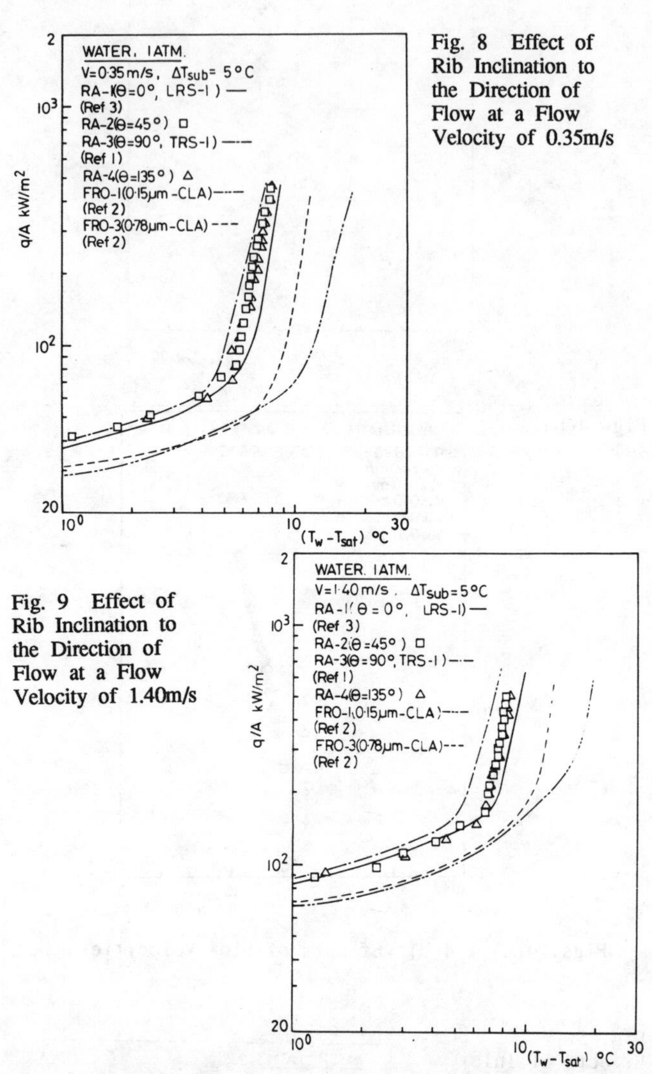

Fig. 8 Effect of Rib Inclination to the Direction of Flow at a Flow Velocity of 0.35m/s

Fig. 9 Effect of Rib Inclination to the Direction of Flow at a Flow Velocity of 1.40m/s

Figs 10(a) and (b) show typical variations of 'h' with 'θ' for a flow velocity of 0.7 m/s and inlet subcoolings of 5°C and 30°C respectively. It is clear from these figures that the specimens 'RA-3' (θ = 90°) and 'RA-1' (θ = 0°) exhibit the highest and lowest heat transfer performances respectively. As expected, the heat transfer performance of the specimens with θ = 45° and 135° were identical and had a value between the two extreme cases.

The experimental data of the heating surfaces employed in this study are correlated, as shown in Fig 11, by the following equation to within ±30%:

$$\frac{Nu_{expt}}{Nu_{calc}} = 114.0 \left[\frac{q/A}{\lambda\rho_v V}\right]^{0.73} \left[\frac{\lambda}{C_{p\ell}\Delta T_{sub}}\right]^{0.55} \left[\frac{\rho_v}{\rho_\ell}\right]^{0.70} (1 + \sin\theta)^{0.15}$$

The physical significance of the terms

$$\left[\frac{q/A}{\lambda\rho_v V}\right] \quad \left[\frac{\lambda}{C_{p\ell}\Delta T_{sub}}\right] \quad \text{and} \quad \left[\frac{\rho_v}{\rho_\ell}\right]$$

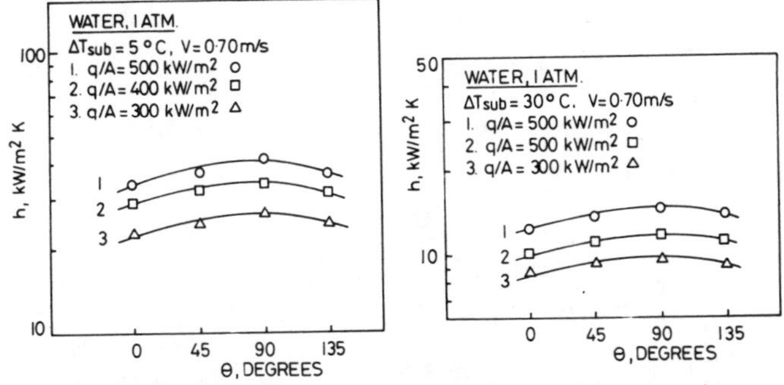

Fig. 10 Typical Plots of 'h' vs 'θ' Showing the Effect of Rib Inclination to the Direction of Flow

is well documented in the literature and the inclusion of the term (1 + sinθ) describes the effect of rib orientation to the direction of flow on the boiling heat transfer coefficient. The single-phase Nusselt number (Nu_{cal}) is determined from the relationship given in James (1967) and was developed for heat transfer in asymmetrically heated rectangular ducts. It correlated the single-phase heat transfer data of this work to within ±10%. The dimensionless groups in the Nu_{expt}/Nu_{cal} relationship were evaluated using computer programs in conjunction with a CDC 7600 series computer and the coefficient and power indices were determined by conventional techniques. Full development of the correlation is given in Akhanda (1985b).

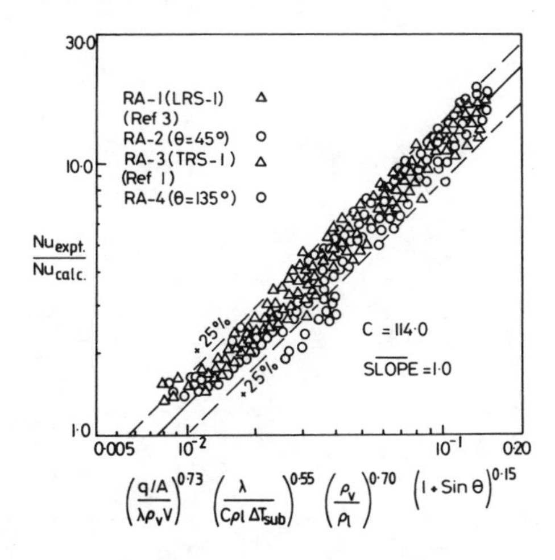

Fig. 11 Correlation of the Boiling Data from 'RA' Surfaces

CONCLUSIONS

1 The effects of rib orientation, ie rib angle (θ) to the direction of flow were investigated experimentally. The effect of surface micro-roughness was isolated and surfaces employed were 'aged' prior to recording the experimental data. Rib length, height and spacing were maintained constant. Augmentation in heat transfer rate was observed up to a value of θ = 90°and further increase in the rib angle (θ) resulted in a decrease in the heat transfer rate. As expected identical results were obtained for rib angle (θ) of 45° and 135° and it was observed that the heat transfer performance was poorest for rib angle θ = 0° i.e. longitudinal ribs.

2 The optimum surface 'RA-3', having θ = 90° showed a substantial improvement in heat transfer performance when compared with the optimum roughened surface for all inlet subcoolings and flow velocities.

3 A relationship is presented which correlates the data to within ±30% over the experimental range investigated.

ACKNOWLEDGEMENT

One of the authors, M A R Akhanda, is greatly indebted to the Commonwealth Scholarship Commission, UK for the award of a Commonwealth Scholarship, held at the Mechanical Engineering Department at UMIST.

REFERENCES

1. Akhanda, M.A.R., and James, D.D., 1984a, "Enhanced Heat Transfer from Artificially Prepared Ribbed Surface in Flow Boiling", *11th Annual Research Meeting on Heat Transfer and Catalysis and Catalytic Reaction*, Institute of Chemical Engineers (IChemE), University of Bath, UK.
2. Akhanda, M.A.R. and James, D.D., 1984b, "Forced Convection Boiling from Roughened Surfaces", *11th Annual Research Meeting on Heat Transfer and Catalysis and Catalytic Reaction*, I Chem E, University of Bath, UK.
3. Akhanda, M.A.R., and James, D.D., 1985a, "An Experimental Study of the Relative Effects of Transverse and Longitudinal Ribbing of the Heat Transfer Surface in Forced Convective Boiling", *23rd National Heat Transfer Conference, ASME*, Denver, USA.
4. Akhanda, M.A.R., 1985b, "Enhanced Heat Transfer in Forced Convective Boiling", PhD Thesis, University of Manchester, UK.
5. Bergles, A.E., Collier, J.G., Delhaye, J.M., Hewitt, G.F. and Mayinger, F., 1981 "Augmentation of Two-Phase Flow Heat Transfer", *Two-Phase Flow and Heat Transfer in Power and Process Industries*, Hemisphere, Washington DC, pp.366-382.
6. Bondurant, D.L. and Westwater, J.W., 1971, "Performance of Transverse Fins for Boiling Heat Transfer", *Chem Eng Prog Symp Ser*, Vol 67, No 113, p.30.
7. Canilova, G. N., 1965, "Influence of Pressure and Temperature on Heat Exchange in Boiling of Halogenerated Hydrocarbons", *Cholodilnaja Technika*, Vol 42, No 2, p.36.
8. Dunn, P.D. and Reay, D. A., 1976, "Heat Pipes", Pergamon Press.
9. Gorenflo, D., 1966, "Zum Warmubergang bei der Blasenverdampfung an Rippenrohren", Diss T H, Karlsruhe, W Germany.
10. Gorenflo, D., 1968, "Zur Druckabhangigkeit des Warmenuberganges an Siedende Kaltemittel bei freier Konvektion", *Chemie-Ingenieur-Technik*, Vol 40, No 15, pp.757-762.
11. Haley, K.W. and Westwater, J.W., 1965a, "Heat Transfer from a Fin to a Boiling Liquid", *Chem Eng Science*, Vol 20, p.711.
12. Haley, K.W., 1965b, "Design of Fins for Boiling Heat Transfer", PhD Dissertation, University of Illinois, USA.
13. Hesse, G., 1973, "Heat Transfer in Nucleate Boiling, Maximum Heat Flux and Transition Boiling", *Int J Heat and Mass Transfer*, Vol 16, p.1611.
14. Hsu, Y.Y., 1968, "Analysis of Boiling on a Fin", NASA, TND4797.
15. James, D.D., 1967, "Forced Convection Heat Transfer in Ducts of Non-Circular Section", PhD Thesis, Imperial College, University of London.
16. Joudi, K.A., 1977, "Surface Geometry Variations in Nucleate Pool Boiling" PhD Thesis, Department of Mechanical Engineering, UMIST, Manchester.
17. Klein, G.J. and Westwater, J.W., 1971, "Heat Transfer from Multiple Spines to Boiling Liquids", *AIChE Journal*, Vol 17, p.1050.
18. Lai, F.S. and Hsu, Y.Y., 1967, "Temperature Distribution in a Fin Partially Cooled by Nucleate Boiling", *AIChE Journal*, Vol 13, No 4, p1317.
19. Simon-Tov, M., 1970, "Analysis and Design of Extended Surfaces in Boiling Liquids" *Chem Eng Prog Symp Ser*, Vol 66, No 102, p.174.
20. Webb, R.L., 1981, "The evolution of Enhanced Surface Geometries for Nucleate Boiling", *J Heat Transfer Engineering*, Vol 2 Nos 3-4, pp 46-67.

METASTABLE (SUPERHEATED LIQUID) CONDITIONS IN FLOW OF R-22 THROUGH COPPER CAPILLARY TUBES

S. J. Kuehl and V. W. Goldschmidt
Ray W. Herrick Laboratories
School of Mechanical Engineering
Purdue University
West Lafayette, Indiana

ABSTRACT

Data are presented for flow of refrigerant R22 through a copper capillary tube with an inside diameter of 1.245 mm (0.049 in.). The data strongly infer the presence of a notable region where the refrigerant reaches a pressure below saturation but remains in a liquid state. A review of the literature indicates contradictory results, but a general agreement on the existence of this metastable region. It is demonstrated that when represented in terms of a dimensionless underpressure (at the point of flashing), akin to a cavitation number and related to the Reynolds number, the dependence on inlet conditions is lost. Furthermore, there is partial data suggesting that the dependence on inside diameter and type of refrigerant may also be masked.

NOMENCLATURE

d internal diameter of capillary tube (in) (mm)
G mass flux (lbm/h in^2) (kg/h m^2)
L tube length (in) (mm)
$\dot{m}$ mass flow rate (lbm/h) (kg/h)
P static pressure (lbf/in^2) (kPa)
P_{inlet} pressure at the inlet to the capillary tube (lbf/in^2) (kPa)
p_u under pressure (measure of liquid superheat in psi) (kPa)
Re Reynolds number
U_m mean fluid velocity over the cross-section (ft/s) (m/s)
μ dynamic viscosity (lbm/ft*h) (kg/m*s)
γ kinamatic viscosity (ft^2/s) (m^2/s)
ρ fluid density (lbm/ft^3) (kg/m^3)

INTRODUCTION

Under normal operating conditions, a capillary tube in a vapor compression cycle acts as a flow controller, and expander. The refrigerant enters in a subcooled liquid state, flashes, and exits in a two phase (choked flow) condition. The simplest approach is to treat the flow as two regimes, single-phase liquid up to the flashing point, and two phase flow beyond that.

Accounting for these regions properly is necessary in modeling for selection and design (see references in Kuehl (1987)). There is of course considerable concern as to the proper selection of friction coefficients in the two-phase flow, and for that matter even in the single-phase liquid flow (Kuehl (1987)).

However, an additional concern to be reckoned with is that in many instances a mestastable region is also found to exist. In this region, the refrigerant is in a liquid state, even though its average pressure is below the saturation value. This metastable region is also referred to as the superheated liquid region. It has been described by either a "delay length" (length along the tube from the point that saturation pressure is reached to the point at which flashing occurs), by an "underpressure" (the difference between the local pressure and the saturation pressure at the point of flashing, or by a "mestastable superheat" (the difference between the local temperature and the saturation temperature at the point of flashing).

Figure 1 idealizes the situation. The flashing point is easily identified from refrigerant temperatures (usually approximated by the tube surface temperatures). The pressures, if not measured can be inferred from the liquid turbulent fully developed flow conditions (based on actual Reynolds number dependent friction factors). While there is some uncertainty in the exact location of the interface test data, such as that exhibited in figure 2, clearly exhibit the location of the flashing point. (Additional plots are presented in Kuehl and Goldschmidt (1990).) Wedekind (1965) presents documentation of the oscillation in the effective points of complete vaporization (in a slightly larger tube). This, in turn, suggests fluctuations in the flashing point.

REPRESENTATION OF METASTABLE REGION

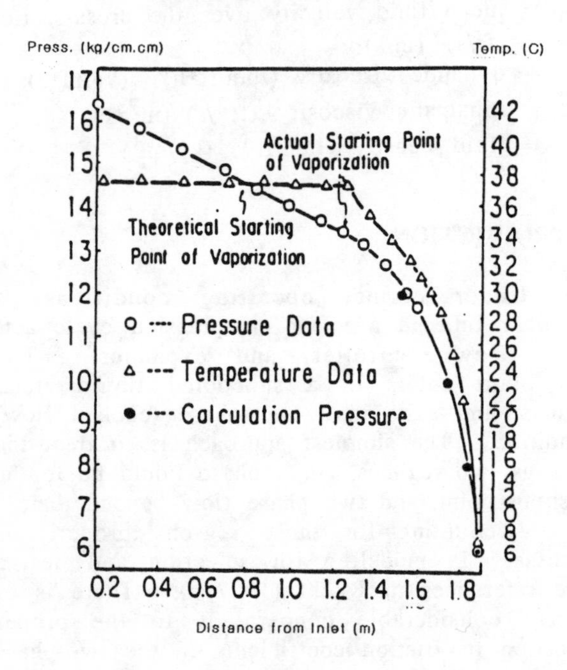

(From Koizumi et al. 1980)

Figure 1: Representation of Metastable Region.

The purpose at this time is to bring in sharper focus the nature of this superheated liquid region through additional test data. The test set up and some data have been presented in Kuehl and Goldschmidt (1990). The reader is referred to that reference and Kuehl (1987) for details.

TEMPERATURE PROFILE
D=0.049", L=31.5" TUBE

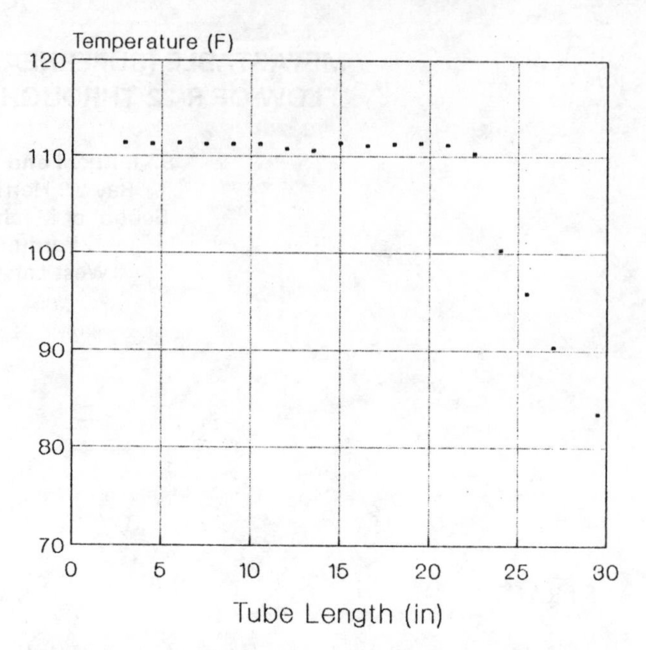

ṁ=96.6 lbm.h; Pi=278.2 psia
Po=119.3 psia; Tsc=9.9F

Figure 2: Temperature Profile for d=0.049 in L=31.5 in capillary tube.

BACKGROUND

Li et al. (1989) and Kuehl and Goldschmidt (1990) present a partial literature review of relevant publications, including some of the concerns with the identification of the metastable region. Hence, a detailed review is not presented at this time, but simply a summary of the salient contributions.

Bolstad and Jordan [1948, 1949] performed the first major study on capillary tube flow. They instrumented a capillary tube with pressure gauges and thermocouples along the tube length; however, they did not identify a metastable region possible due to the uncertainty in their data, as the thermocouples were place several inches apart thus making the temperature profile somewhat smoother than reality.

Cooper et al. (1957) may have been the first to witness metastable superheated liquid flow. Their results were that: delay length increases with decreasing internal diameter; with increasing tube length; with increasing inlet pressure; and with increasing inlet subcooling or in essence the delay in vaporization increases with increasing pressure drop per unit length in the subcooled liquid region.

Mikol (1963), and Mikol and Dudley (1964) presented results of visualization in a glass tube and detailed data in a copper tube. They found metastable flow under all of their test cases, and attempted to quantify the vapor inception point in terms of cavitation parameters. This seems odd because their observations showed that the first bubbles nucleated along the tube wall, not in the fluid core. However, Mikol and Dudley believed that nucleation was brought on by local turbulent pressure fluctuations in the liquid stream. An attempt was made to correlate bubble nucleation superheat with subcooling exhibiting a genual trend of increasing superheat with subcooling. (See also Scott (1976)). Nucleation superheats from 0.1 to about 15°F were noted as the inlet subcooling ranged from 0.1 to 30°F. Daily (1964) suggested that the superheat is necessary to overcome surface tension effects inhibiting the formation of bubbles during boiling, explaining the underpressure as a liquid tension.

In spite of the work of Cooper et al., Mikol & Dudley and others such as Niaz and de Vahl Davis (1969) there are those (such as Erth (1969)) that have suggested that the metastable flow regime is an anomaly resulting in unique laboratory conditions.

Other researchers in the 1980's have also demonstrated the existence of metastable flows. Amongst them, Koizumi & Yokoyama (1980) found liquid superheat pressures of (138 to 207 kPa) 20 to 30 psi and proposed that the underpressure increases with flow and with a decrease of the inside diameter; Rezk & Awn (1979) presented data exhibiting metastable flow regions; while Maczek et al. (1983) addressed the metastable refrigerant flow phenomena with a two-phase model based on the creation and expansion of nucleate bubbles inside the superheated liquid (metastable region). Kuipjers and Janssen (1983), following numerous tests correlated under pressure (liquid superheat with mass flux and inlet subcooling. They suggested that turbulence plays an important role in the magnitude of the under pressure, given by

$$\delta P_{inception} = \delta P_{inception,\ static} - \psi \frac{G^2}{2\rho_f}$$

where the factor ψ takes into account the three-dimensionality of the velocity vector. Kuipjers and Janssen's attempts to correlate underpressure proved to be only mildly successful. However, a trend of decreasing underpressure with increasing mass flux was noted while the superheat first increased then decreased with inlet temperature.

Interestingly, Pate and Tree (1984) while concentrating on capillary tube with heat transfer, concluded - based on extremely carefully conducted and detailed measurements - that there is a metastable region for the adiabatic capillary tube, but none for the tube with heat transfer. They explain this in part by the effects of the heat transfer on the temperature distribution of the refrigerant. Finally, Li et al. (1989) also based on detailed data, suggest that in addition to the metastable liquid flow region there also is a metastable two phase flow region. Their data suggest an increasing under pressure with mass flow rate, but decreasing with inlet subcooling, and their data also suggest that the increase in underpressure is attributed to an increase in the depressurization rate, while turbulent pressure fluctuations would tend to decrease the underpressure.

In sum, the results available to date lead to the conclusion that there indeed is a region of metastable flow. It appears to possibly depend on inside diameter, rate of depressurization, mass flow rate, pressure fluctuations, surface tension, heat transfer, and inlet subcooling as well as, obviously, the surface finish of the tube.

The controversy on whether or not a metastable region does occur is settled by the consistent results reported. However, what is not settled is what this depends on. Furthermore, there is a shortage of data based on R22, which is now in part filled.

TEST RESULTS

The test configuration is described in Kuehl and Goldschmidt (1990) and in Kuehl (1987). Hence, it is not detailed at this time. It consisted of a closed flow set up, driven by a conventional unitary heat pump compressor, with an oil separator and filter upstream of an instrumented capillary tube. The instrumentation consisted of inlet and outlet pressures and temperatures, plus temperatures along the tube, and mass flow rate measurements. The uncertainties of these were nominally under 1% for the pressure readings (measured with a Vernitech transducer), ± 0.2C (±0.3°F) for the temperatures (measured with type K thermocouples) and ± 0.01 kg/min (± 0.025 lbm/min) for the mass flow rates (measured with a Micro-motion coriolis type meter).

Only one capillary tube was used in the testing. It had an inside diameter of 1.245 mm (0.049 in) and the length of 0.8 m (31.5 in). The location of the flashing point was determined from the temperature profiles (sample curve was given in figure 2), while the pressure distribution was determined from the liquid flow friction factors and measured mass flow rates.

Figure 3 shows the "vaporization delay lengths". These lengths were determined by subtracting the equilibrium liquid length, from the experimental liquid length. The experimental length was determined by noting where the temperature profile for the capillary tube began to decay along the tube length. The uncertainty of this length is no better than ± 3.8 cm (± 1.5 in) due to the spacing of the thermocouples soldered to the capillary tube surface and the uncertainty in the temperature measurement. (The circled data points were suspect of being below critical flow condition).

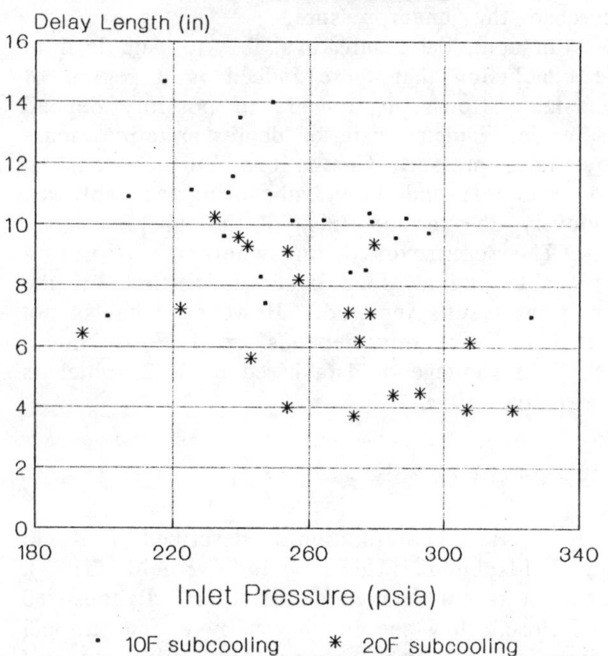

Figure 3: Delay lengths vs. Inlet Pressure

From the data, it was observed that certain tube locations served as preferred flashing points for small ranges of temperatures and pressures. Quite possibly these locations are favorable bubble nucleation sites in the surface of the tube bore. These sites would be entirely random in location and affected by the manufacturing process.

Examination of the plots in Figure 3 suggests that:

1. Delay length decreases with an increase in subcooling.
2. Delay length decreases with an increase in inlet pressure.

This is opposite to the observations of Cooper et al. (1953) plus Koizumi and Yokoyama (1980), but in agreement with Li (1989).

The dependence of underpressure on mass flux is shown on figure 4. (Again, the two data points suspect of having flow below the critical value are circled). While there is considerable scatter in the data, it does show a consistent increase of underpressure with mass flux. As a matter of fact, forcing an extrapolation to the intercept would suggest that the increase could be parabolic. This is in agreement with the results of Koizumi (1980) and Li (1989), but in disagreement with the results of Kuipjers (1983). The plots also show a dependence on subcooling, with a higher underpressure for lower values of subcooling.

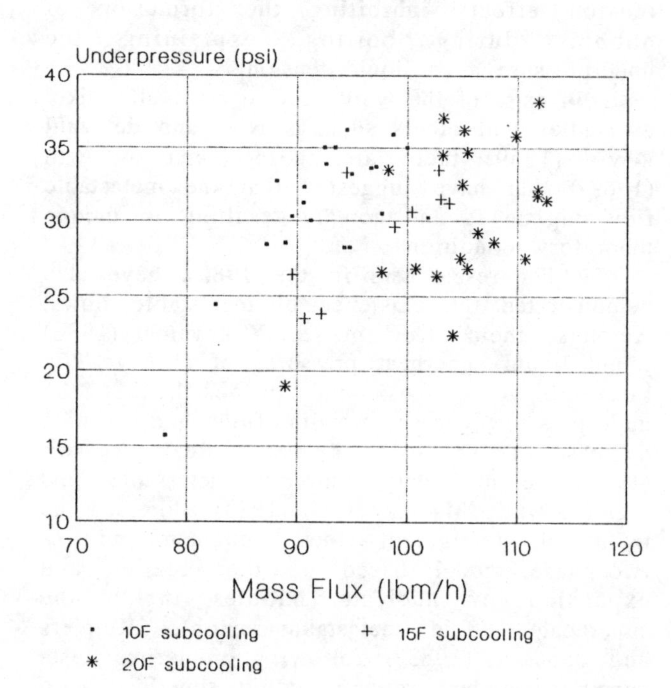

Figure 4: Underpressure vs. mass flux

DISCUSSION

Numerous attempts at correlating the dependence of underpressure on other parameters have had, at best, partial success. While Mikol and Dudley suggested a possible dependence on Weber number most other researchers considered the relationship between nondimensionless parameters - making the results shy of generalization. Any expectations for generalization must start with dimensionless parameters.

Figure 5 plots a liquid superheat number or dimensionless underpressure $\frac{p_u}{\frac{1}{2}\rho U_m^2}$ (in terms similar to the Euler Number or the Cavitation Index) versus the Reynolds number. (The properties at the inlet are used in the analysis). Accounting for the experimental uncertainty shows these data to essentially fall on one common trend for the three levels of superheat. The Reynolds numbers for the tests are definitely in the turbulent regime. Li et al. (1989) present sufficient data to reduce two of their points to the same dimensionless groups. (These are obtained from their figures 3 and 7 and 8; sufficient data are not given to determine the grouping for the other runs as property values at the inlet are needed.) Their Reynolds numbers appear to also be in the turbulent region. On the other hand, the data of Kuipjers and Janssen (1983) exhibit Reynolds numbers below and near the transition region (for larger size pipes). There is hence no assurance that their flow was indeed turbulent, hence the behavior of their data may be different from that of others that were definitively in the turbulent regime. (Different trends were noted in the mass flow dependence and subcooling dependence of the underpressure).

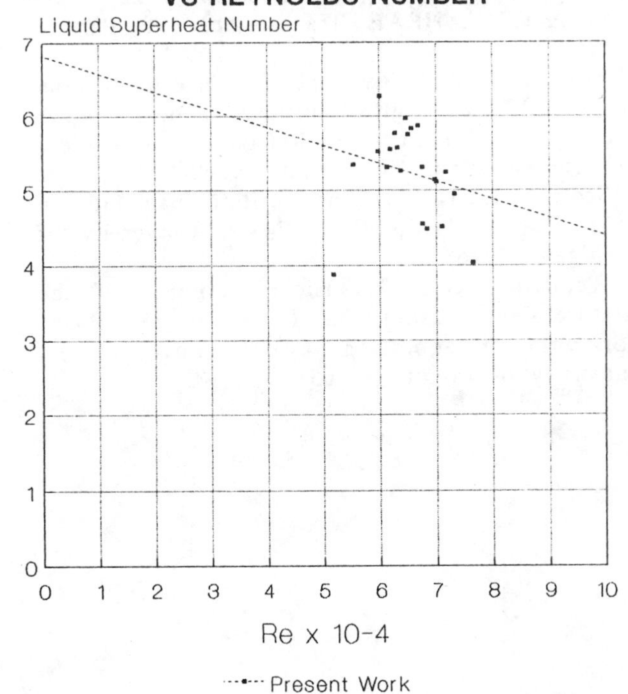

Figure 5: Dimensionless Underpressure (Liquid Superheat Number) vs. Reynolds Number.

Figure 6 compares the present results with those obtained from Li et al., Mikol and co-workers and Kuipjers and Janssen. Some of these values were obtained estimating properties and other quantities from the data provided.

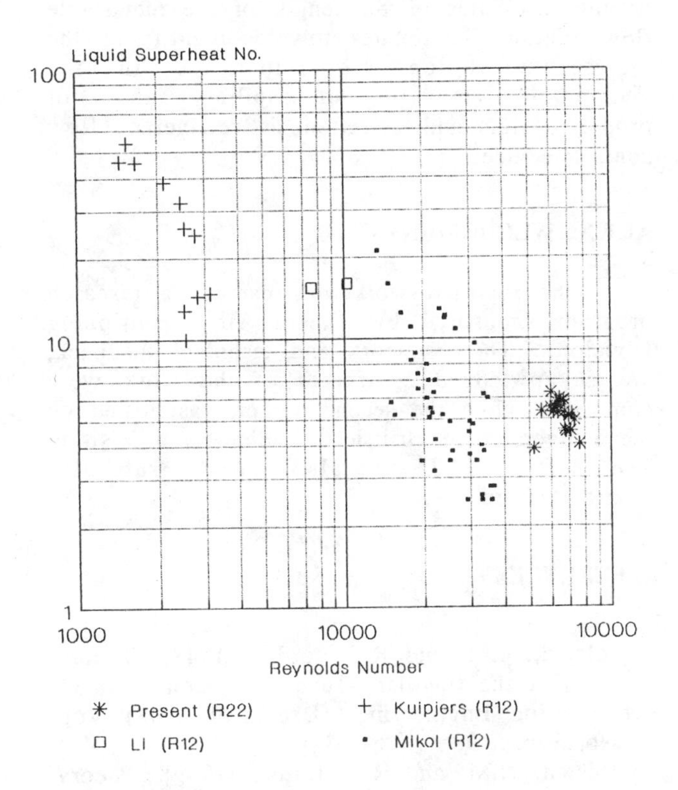

Figure 6: Comparison with other Results.

It should not be surprising that the liquid superheat number goes to infinity as the Reynolds number decreases. This is forced by two conditions. Firstly, in the absence of flow there would be no frictional pressure drop and hence decrease of the subcooling towards flashing. Furthermore, the liquid superheat number is made dimensionless with a velocity squared in the denominator. What is surprising though is that both sets of data, for R12 and R22, appear to follow somewhat of a similar curve. More data are indeed needed at lower Reynolds number for the R22 and higher Reynolds number for the R12 to determine whether there might be a dependence on surface tension coefficients, or on the depressurization rate as has been suggested by earlier investigators.

CONCLUSIONS

Adiabatic refrigerant flow in capillary tubes, as used in vapor compression cycles, exhibits a metastable flow of superheated liquid.

Measurements taken at different inlet conditions provide a general dependence of the underpressure made dimensionless by the dynamic pressure (liquid superheat number) on Reynolds Number. These, as generalized, can provide estimates of the length of the metastable flow region. Estimates toward quantifying the metastable region are imperative in the mathematical modeling of capillary tubes for proper design and selection as refrigerant flow control devices.

ACKNOWLEDGMENTS

The reported work was part of a research program sponsored by Rheem Air Conditioning Division. Their support, and technical feedback, are acknowledged. Aspects of the work were completed while the second author was hosted by the University of British Columbia as a Visiting Professor. They also are gratefully acknowledged.

REFERENCES

Bolstad, N.M. and R.C. Jordan [1948] "Theory and Use of the Capillary Tube Expansion Device," Refrigerating Engineering, (December 1948) Vol. 56, No. 6, p. 519.

Bolstad, N.M, and R.C. Jordan [1949] "Theory and Use of the Capillary Tube Expansion Device, Part II, Nonadiabatic Flow," Refrigerating Engineering, (June 1949), pp. 572-583.

Cooper, L., Chu, C.K. and W.R. Brisken [1957] "Simple Selection Method for Capillaries Derived from Physical Flow Conditions," Refrigerating Engineering, Vol. 65, No. 7 (July 1957), p. 37.

Daily, J.M., Discussion on Mikol & Dudley's paper, "Transactions of the ASME, Series D; Journal of Basic Engineering (June 1964) pp. 261-262.

Erth, R.A., [1969] "Two Phase Flow in Refrigeration Capillary Tubes: Analysis and Prediction," PhD. Thesis, Purdue University (September 26, 1969).

Koizumi H. and K. Yokoyama, [1980] "Characteristics of Refrigerant Flow in a Capillary Tube," ASHRAE Transactions 86, Part 2, (1980), pp. 19-27.

Kuehl, S. J. "Study, Validation and Improvement of a Model for Sizing a Capillary Tube for an Air-Conditioning/Heat Pump System" MSME thesis, Purdue University, August 1987.

Kuehl, S. J. and Goldschmidt, V. W. "Steady flows of R22 through capillary tubes: Test Data", ASHRAE Transactions, Vol. 96, Pt.1, 1990.

Kuipjers, L.J.M. and M.J.P. Janssen [1983] "Influence of Thermal Nonequilibrium on Capillary Tube Mass Flow," Proceedings of the XVIth International Congress of Refrigeration, Commission B2. (Paris - 1983) pp. 307-315.

Li, Rui-Lang, Sui Lin, Zu-Yao Chen and Zhi-Hang Chen "Metastable Flow of Refrigerant-12 through Capillary Tubes" Private Communication, August 1989; paper submitted for publication to the IJR, 1989.

Maczek, K. Krolicki, Z. and E. Sochanecka, [1983] "Model of Throttling Capillary Tube and Metastable Process," Proceedings of the XVIth International Congress of Refrigeration, Commission B2. (Paris -1983) pp. 154-161.

Mikol, E.P., [1963] "Adiabatic Single and Two-Phase Flow in Small Bore Tubes," ASHRAE Journal, Vol. 5, No. 11, pp. 75-86.

Mikol, E.P. and J.C. Dudley, [1964] "Visual and Photographic Study of the Inception of Vaporization in Adiabatic Flow," Transactions of the ASME, Series D: Journal of Basic Engineering, (June - 1964), pp. 257-264.

Niaz, R.H. and G. de Vahl Davis, [1969] "Adiabatic Two-Phase Flow in Capillary Tube," Cocurrent Gas-Liquid Flow, Symposium of the Canadian Society of Chemical Engineers, (University of Waterloo, Canada - 1969,) Vol. 1, pp. 259-269.

Pate, M.B. and D. R. Tree, [1984] "An Analysis of Pressure and Temperature Measurements Along a Capillary Tube-Suction Line Heat Exchanger," ASHRAE Transactions, 90 Part 2 (1984).

Rezk, A. and A. Awn, [1979] "Investigation on Flow R-12 Through Capillary Tubes," XVth International Congress of Refrigeration (Venice-1979), Proceedings Vol. II pp. 443-452.

Scott, T. C., [1976] "Flashing Refrigerant Flow in Small Bore Tubes", PhD. Thesis, University of Michigan (1976).

Wedekind, G.L. "Transient Response of the Mixture-Vapor Transition Point in Two-Phase Horizontal Evaporating Flow", PhD Thesis, University of Illinois, (1965).

CONDENSATION OF THE STEAM/NONANE SYSTEM ON A VERTICAL FLAT PLATE AND A SINGLE COLUMN OF HORIZONTAL TUBES

H. J. Hoon and B. M. Burnside
Department of Mechanical Engineering
Heriot-Watt University
Edinburgh, Scotland

ABSTRACT

Both azeotropic and nonane rich vapours of the steam/n-nonane system were condensed at atmospheric pressure over a copper vertical flat plate and a single column of six horizontal copper tubes. Over the whole range of temperature driving forces channelling flow of condensate occurred on the oxidised surface and standing drop film flow on the same surface free of oxide. Continuous tests were run for several weeks with the surface in both conditions before collecting data. The results are compared with a generalised channelling flow model and a simple model of standing drop film condensation.

NOMENCLATURE

a	area fraction
A	area,m^2
c_p	liquid specific heat,kJ/kgK
D	diameter,m
$f,\overline{f}$	mass fraction vapour condensing;mean value
$F_v,\overline{F_v}$	vapour phase mass transfer coefficient;mean
g	gravitational acceleration,m/s^2
$G,\overline{G}$	mass velocity; mean value,$kg/m^2 s$
h_{fg}	latent heat,kJ/kgK
h'_{fg}	h_{fg} + 0.68 $c_p \Delta T_f$,kJ/kgK
H_{fg}	molar latent heat,kJ/kgmolK
k	thermal conductivity,kW/mK
L	length of surface,m
m_z	mass flowrate at position z,kg/s
$M,\overline{M}$	molecular weight;mean value
n	number of tubes in column
N_d	no. drops departing/m^2
Nu	Nusselt number
q	heat flux density,kW/m^2
r	radius,m
r_a	radius after coalescence,m
r_b	radius before coalescence,m
r_d	radius at departure,m
R_m	mean departure radius,m
Re	Reynolds number
Sc	Schmidt number diffusion film
t	time,s
T	temperature,C,K
U,U_z	velocity,m/s
W	surface width,m
x	mass fraction of liquid
X,Y	Mole fraction liquid/vapour
y	mass fraction vapour phase
Z	Droplet number

Greek Letters

α,α_f	film heat transfer coeff,kW/m K
β,γ	constants in equation (21)
δ,δ_z	film thickness; δ at z,m
Δ	finite difference
$\mu,\overline{\mu}$	dynamic viscosity,kg/ms
ν	volume fraction
ρ	density,kg/m^3
σ	surface tension,N/m

Subscripts

A	azeotrope
B	shared surface model
c	condensate
D	Deakin model
f	film
G	General channelling flow model
h	hydraulic mean
i	fluid, 1-water; 2-nonane
min	minimum
n	coalescence number
sat	saturation state
v	vapour
w	water,wall

1. INTRODUCTION

Condensation of mixtures of steam and vapours of organic liquids immiscible with water is a common feature of steam distillation, solvent drying and heterogeneous azeotropic extraction processes. It has been the subject of many studies since the pioneering work of Kirkbride(1934) on steam systems condensing benzene and naptha on a horizontal tube. Results of these studies and theoretical models to predict the condensing heat transfer coefficient,α_f as a function of temperature driving force ΔT_f have been reviewed by

Bernhardt et al.(1971), Boyes and Ponter(1972), Deakin (1976) and Polley(1976).

Three types of condensate flow pattern dominate-channelling, standing drop film and film drop flows. Heat transfer coefficients are fundamentally affected by the pattern of condensate flow (Polley and Callus,1978). They are higher in channelling flow than in the other modes. Bernhardt et al.(1971) have shown that the shared surface model (equations (1) & (10) below) fits a wide range of experimental data including eleven organic liquids with water, condensate water volume fractions between 1.5 and 29%, eleven geometries and four different metal surfaces, to within ±20%. The model is based on Nusselt condensation of each phase on an area proportional to its volume fraction in the condensate. Deakin(1976) has modified the model to allow for different thicknesses of the films. Deakin and Polley(1976) have developed models to predict α_f for the standing drop film mode. The former model requires knowledge of the fractional area occupied by the drops. Based on observed area fractions it predicted α_f for true standing drop film condensation reasonably well. Polley's model is extremely complicated to apply. In this paper a simplified version is developed.

Much doubt exists about whether standing drop film or channelling flow will occur on a particular condensing surface. Polley and Callus(1978) contended that on oxidised surfaces the former mode occurs at low ΔT_f, changing to the latter as ΔT_f increases. Under these circumstances the variation of α_f with ΔT_f depends on the point at which transition occurs. Clearly this varies with the type of surface or degree of oxidation and explains somewhat contradictory α_f v. ΔT_f trends reported (Polley and Callus,1978, Deakin, 1976, Boyes and Ponter,1972).

The work described here (Hoon, 1981) is aimed at clarifying this confusion. The steam/nonane system was used, condensing on a copper vertical plate and on a column of six horizontal tubes. Both the azeotrope and nonane rich mixtures were condensed.

2. THEORETICAL MODELS

General Channelling Flow Model.

The two liquids are assumed to flow separately down a flat surface as Nusselt laminar films with condensation of each phase on its own channel and with common $\Delta T_f = T_A - T_W$. The overall film coefficient can then be written

$$\alpha_f = a\,\alpha_1 + (1-a)\,\alpha_2 \qquad (1)$$

The mass flowrates of liquid at distance z from the top of the surface are

$$m_{z1} = \rho_1 U_{z1} \delta_{z1} W_1 \qquad m_{z2} = \rho_2 U_{z2} \delta_{z2} W_2 \qquad (2)$$

The condensate mass fraction of phase 1 on the surface is

$$x_{1c} = m_{1c}/(m_{1c}+m_{2c}) = 1/(1+(\rho_2/\rho_1)(U_1/U_2)(\delta_2/\delta_1)(W_2/W_1)) \qquad (3)$$

Since $W_1/W_2 = a/(1-a)$

equation (3) becomes

$$a_1 = 1/(1+(1-x_{1c})/x_{1c}\,(\rho_1/\rho_2)(U_1/U_2)(\delta_1/\delta_2)) \qquad (4)$$

and changing to condensate volume fraction this becomes

$$a_1 = 1/(1+(1-\nu_{1c})/\nu_{1c}\,(U_1/U_2)(\delta_1/\delta_2)) \qquad (5)$$

Writing equation (5) in terms of properties gives

$$a = 1/(1+(1-x_{1c})/x_{1c}\,((\rho_1/\rho_2)^2 (k_1/k_2)^3 (\mu_2/\mu_1) \times (h'_{fg2}/h'_{fg1})^3))^{1/4} \qquad (6)$$

The mean α_f for the surface can be obtained using equations (1),(6) and either

$$\alpha_i = 0.943(g\rho_i^2 k_i^3 h'_{fgi}/(L\mu_i \Delta T_f))^{\frac{1}{4}} \qquad (7)$$

for a vertical surface, length L, or

$$\alpha_i = 0.728(g\rho_i^2 k_i^3 h'_{fgi}/(nD\mu_i \Delta T_f))^{\frac{1}{4}} \qquad (8)$$

for a vertical column of n horizontal tubes,diameter D. If $U_1 = U_2$ equation (5) becomes

$$a_{1D} = 1/(1+(1-\nu_{1c})/\nu_{1c}\,(\delta_1/\delta_2)) \qquad (9)$$

which is the area fraction of Deakin's model. If $U_1 = U_2$ and $\delta_1 = \delta_2$ then

$$a_{1B} = \nu_{1c} \qquad (10)$$

which is the Bernhardt et al.(1971) shared surface model. It should be noted that the Deakin and shared surface models do not satisfy energy conservation, whereas the general model, equation (6) does. The three models are compared in table 1 for a range of steam/organic systems.

Systems Steam with	Ratios	ΔT_f (degC) 5	30
benzene	α_B/α_G	0.952	0.939
	α_D/α_G	0.960	0.952
toluene	α_B/α_G	0.936	0.943
	α_D/α_G	0.967	0.963
trichloroethylene	α_B/α_G	0.963	0.950
	α_D/α_G	0.937	0.919
n-heptane	α_B/α_G	0.862	0.870
	α_D/α_G	0.906	0.901
R113	α_B/α_G	0.972	0.971
	α_D/α_G	0.962	0.984
p-xylene	α_B/α_G	0.955	0.949
	α_D/α_G	1.000	0.984
n-nonane	α_B/α_G	0.876	0.882
	α_D/α_G	0.982	0.975

Table 1. Heat transfer coefficients calculated by Shared Surface,Deakin and General channelling flow condensation models.

The shared surface model predicts α_f values generally up to 7% below the new model values but underpredicts by 12% for steam/nonane and 14% for steam/heptane. The Deakin model places α_f intermediate between the other two models.

Standing Drop Film Model.

This is a simplified version of the Polley(1976) model. The main assumptions (Hoon, 1981) are:-

(a) The surface is divided into separate film flow and standing drop regions. Only organic condenses on the film region forming a Nusselt film with wavy flow on its surface.

(b) Steam alone condenses on the standing drop region, the water forming hemispherical standing drops, uniformly distributed and of uniform size, which grow by coalescence with neighbouring drops of the same generation to the departure size. They then slide down the surface. In this model only two drops were assumed to take part in each coalescence, compared to four in the Polley model, since it was observed that the actual droplet size distribution is not uniform, figure 1b. For the same reason a mean droplet departure radius, R_m was used instead of the maximum departure size used by Polley(1976). A value of R_m=0.75mm was found to give the best fit to the experimental data. This compares with values of droplet departure radii between 0.35 and 0.60mm observed in dropwise condensation of steam (Glicksman and Hunt, 1972; Rose and Glicksman, 1973; Graham and Griffith, 1973).

(c) Both regions have the same temperature driving force and degree of subcooling.

(d) The sweeping effect of condensate flow and mass transfer resistance at the liquid/vapour interface are neglected.

Analysis.

The conduction heat transfer coefficient, α, through a single hemispherical water droplet of radius, r, is $\alpha = 4k_1/r$ and the rate of heat transfer through the drop, Q, is also equal to the latent heat released

$$Q = 4\pi r k_1 \Delta T_f = \rho_1 h'_{fg1} \times d(2\pi r^3/3)/dt \qquad (11)$$

If Z_o is the initial number of drops/unit area and uniform droplet size and spacing assumed, the first coalescence occurs when the droplet radius is $r_{b1} = 1/(2Z_o^{1/2})$. Just after coalescence of two droplets $r_{a1} = 2^{1/3} r_{b1}$ and the number of droplets is $Z_1 = Z_o/2$. The initial equilibrium droplet radius is given by $r_{min} = 2T_{sat}\sigma/(h_{fg1}\rho_1 \Delta T_f)$.

The radii before and after the nth coalescence are given by

$$r_{bn} = 1/(2Z_{n-1}^{1/2}) = 1/2(2^{n-1}/Z_o)^{\frac{1}{2}} \qquad (12)$$

and $$r_{an} = 2^{\frac{1}{3}} r_{bn} \qquad (13)$$

The number of coalesced droplets is $Z_n = Z_{n-1}/2 = Z_o/2^n$ and the time interval between coalescences, obtained by rearranging and integrating equation (11), is

$$\Delta t_n = \rho_1 h'_{fg1} \; (r_{bn}^2 - r_{a\,n-1}^2)/(4k_1 \Delta T_f) \qquad (14)$$

The drops grow from $r_{a\,n-1}$ to r_{bn} by condensation and from r_{bn} to r_{an} by coalescence. The process is repeated until the departure size $r_d = R_m$ is reached. A new growth cycle begins immediately. The number of drops/unit area at departure, N_d, radius r_d, and the cycle time, t_c, which is the sum of the intervals Δt_n, equation (14), are now known. The rate of heat transfer/unit area, q, to the standing drop region is therefore

$$q = (2/3)\pi r_d^3 \rho_1 N_d \; h'_{fg1} \; /t_c \qquad (15)$$

If the mean departure radius, R_m, lies between the drop sizes before and after the nth coalescence, then $r_d = r_{an}$, otherwise $r_d = R_m$.

The rate of heat transfer/unit area, q_2, through the laminar wavy film region, area A_2, is based on the Nusselt film coefficient, α_2, given by equations (7) or (8), with the Labuntsov(1957) correction

$$q_2 = \alpha'_2 \Delta T_f \; (0.95 \; Re_{f2}^{0.04}) \qquad (16)$$

The overall time averaged heat transfer coefficient, α_f, is therefore

$$\alpha_f = (a_1 \, q_1 + (1-a_1) \, q_2)/\Delta T_f \qquad (17)$$

where

$$a = 1/(1 + (1-x_{1c})/x_{1c}) \; (q_1/q_2) \; (h'_{fg2}/h'_{fg1})) \qquad (18)$$

The film Reynolds number can be written in the form

$$Re_{f2} = (3.8\alpha'_2 \; \Delta T_f \; (A_2/W_2))^{1.04}/(\mu_2 h'_{fg2}) \qquad (19)$$

The values of (A_2/W_2) for a vertical surface height L, a horizontal tube diameter D and a vertical column of n horizontal tubes are L, πD and nπD, respectively. The initial area density of drops, Z_o, was assumed to be $10^7/cm^2$. The results were insensitive to this value. Finally, the mean heat transfer coefficient for the surface can be calculated using equation (19), making use of equations (15), (16), (18) and (19), (Hoon,1981).

Compared to the model developed above, the Polley model is more realistic in assuming two phase condensation on each region. However, Polley subsequently simplified the model by assuming that the organic liquid, condensing in the standing drop film region, returns immediately to the film region and does not affect heat and mass transfers in the former region. The Polley model is very complicated and requires a local heat and mass transfer analysis from location to location. Futher it cannot be easily adapted to model condensation of non-azeotropic mixtures. This problem does not arise in the present model since the condensate composition is directly related to the heat transfer in each region.

Non-azeotropic Vapour- Condensate Composition.

When the vapour is rich in one of the components the proportion of the rich component in the condensate is not equal to that in the vapour, due to diffusion resistance close to the interface. In this work a modified Colburn and Drew mass and heat diffusion film model was used to calculate the condensate composition X_c. The following assumptions were made:-

(a) The vapour/liquid interface temperature and vapour composition are azeotropic, $Y_1 = Y_A$.

(b) Sensible heat flux to the interface was ignored

(c) Bulk vapour properties, composition, Y, vapour diffusion layer mass transfer coefficient, F_v, and flow parameters were taken to be average values for flow past the surface.

Thus, heat flux through the interface is

$$q = \overline{F}_v \; H_{fgc} \; \ln((X_c - Y_A)/(X_c - \overline{Y}_v)) \qquad (20)$$

Making use of the Chilton and Colburn analogy (Bird et al.,1961)

$$\overline{F}_v = \beta \; Re_v^{\gamma} \; (\overline{G}/\overline{M}) \; (1/\overline{Sc}_v)^{\frac{2}{3}} \qquad (21)$$

where β and γ are constants in the analagous heat transfer correlation

$$Nu = \beta \; Re^{\gamma}$$

The condensation heat flux through the interface, area A_c, is

$$q = m_c \; H_{fgc}/(A_c/M_c) \qquad (22)$$

Using equations (21) and (22) and defining the mass fraction of the mean bulk vapour flow which condenses, $\overline{f}$, as $\overline{f} = m_c/\overline{m}_v$, equation (20) may be written

$$m_c^{\gamma} \; \ln((X_{1c} - Y_{1A})/(X_{1c} - \overline{Y}_{1v})) = \overline{f}^{\gamma+1} \times ((\overline{\mu}_v D_h/4)^{\gamma}/\beta)(\overline{M}/M_c)(A_v/A_c)(\overline{Sc}_v)^{\frac{2}{3}} \qquad (23)$$

where D_h is the hydraulic diameter of the vapour flow

cross section, area, A_v.

It may be shown (Hoon, 1981) that $\overline{Y}_{lv}$ and $\overline{f}$ are given by

$$\overline{Y}_{lv} = (2Y_{lv} - f\ (Y_{lv} + X_{lc}))/(2(1-f)) \qquad (24)$$

and

$$\overline{f} = 2f/(2-f) \qquad (25)$$

The total condensate mass flowrate is a function of condensate composition

$$m_c = \alpha_f\ A_c\ \Delta T_f / (x_{lc}\ h'_{fg1} + (1-x_{lc})\ h'_{fg2}) \qquad (26)$$

Condensing film heat transfer coefficient, α_f, is a function of condensate composition and can be estimated by either the channelling or standing drop film models.

Finally, the condensate mass fraction, x_{lc}, may be calculated by solving equations (23-26). In this study the fraction of the vapour entering the condenser which condensed, lay between 1/4 at low and 1/2 at high condensation rates. An average value, f = 0.375 was used. The best fit to the data for the vertical surface was obtained setting β=462, γ=-0.86, and for the tube column, β=388 and γ=-0.85.

3. APPARATUS

The test rig has been described fully by Hoon(1981). Distilled water and 99.5% pure n-nonane were used in the tests. Contamination was avoided by using only AISI 316 stainless steel, copper, hard nickel plated mild steel and brass fittings, Tempex and Pyrex glass and PTFE and Viton seals in the construction. Separate 4kW boilers were used to allow generation of nonane rich vapours. The vapours were mixed in a mixer box and desuperheated in a partial condenser before passing through a droplet separator into the test condenser. Vapour velocities in both test condensers were never greater than 1.5 m/s. Excess vapour was condensed completely in a Pyrex total condenser. Both test and total condensers were vented by means of a vacuum pump and protected by a liquid nitrogen trap. System pressure was measured by a pressure transducer connected to the test condenser. The dew point inlet vapour temperature was measured by a thermocouple at the condenser inlet.

Condensate from test and total condensers was separated by gravity into water and nonane streams. Flowrates were determined by volume/timing measurements without interrupting feed return to the boilers. From this data the composition of condenser vapour inlet and condensate was calculated. The rig was wrapped in a thick layer of fibreglass insulation.

Vertical Flat Plate Condenser, figure 2.

Vapour entered at the top of the shell and left at the bottom. Condensate was collected in a tray just below the front surface of a 150mm long x 70mm wide x 76mm deep oxygen free high conductivity copper block. Cooling was provided by water flow through a cooling jacket soldered to the back of the block with thermocouples fitted at inlet and outlet. The block was fitted with nine 1mm o.d. mineral insulated metal clad thermocouples in three rows at depths of 20,42 and 64mm from the condensing surface. The temperature of the latter was taken to be the average of the three values extrapolated to the surface. Windows at the front and side of the shell permitted observation of the condensing surface. The flow cross sectional area was 79.2 cm^2 with a hydraulic diameter of 55mm.

Horizontal Tube Column Condenser, figure 3.

The tube column consisted of six 25.4mm o.d. x 6.35mm i.d horizontal high conductivity thick copper tubes, 87mm long, in a vertical column symmetrically disposed in the shell and visible through a window at the front. Vapour entered at the top of the shell and left at the bottom. Condensate was collected on a tray fitted below the tubes. The vapour flow cross sectional area was 90 cm^2 with hydraulic diameter of 62mm. Contact area between the ends of the tubes and the condenser housing was kept small to minimise heat losses. One thermocouple measured the common inlet cooling water temperature. Each tube was fitted with a thermocouple and needle valve at the outlet end. Two 0.5mm diameter mineral insulated metal clad thermocouples were fitted on a common radius at an angle of 120° fromthe top of each tube, with the centre of the outer thermocouple 1.3mm below the surface, figure 3.

4. PRELIMINARY PURE FLUID CONDENSATION TESTS

To check correct functioning of the rig and leak tightness pure steam and pure nonane tests were conducted using the vertical flat surface configuration. Before data was collected in the steam tests, the rig was run continuously for four weeks until only filmwise condensation occurred on a stable oxidised surface. Subsequent condensation of pure nonane on this surface was of course also filmwise. The maximum film Reynolds numbers of water and nonane condensates were 110 and 350, both in the laminar flow regime.

In the steam tests at 1.06 bar, average film heat transfer coefficients, α_f, between 16.5kW/m^2K at ΔT_f=2.7degC and 11.7kW/m^2K at 10.6degC were obtained. These values are higher than the Nusselt predictions, equation (8), of 14.5 and 10.2kW/m^2K but compare well with the values corrected for wavy flow, equation (18), 16.0 and 11.7kW/m^2K respectively. Condensing nonane at 0.53 bar (T_{sat}=127C) the average measured α_f =10.7kW/m^2K at ΔT_f=51.7degC compared with the Nusselt value of 9.4kW/m^2K and the value corrected for wavy flow, 11.0kW/m^2K. These results showed satisfactory venting and leak tightness of the rig.

Surface Preparation.

The experimental procedure is detailed elsewhere (Hoon,1981). After the pure fluid tests using the vertical surface and preliminary runs on the tube configuration, the condensing surfaces were covered with an oxide layer. An unsuspected and previously unreported phenomenon occurred during condensation of steam/nonane mixtures in the oxygen free conditions of the tests (Hoon,1981). The oxide layer on the condensing surface fractured progressively and washed away completely after about 2-3 weeks continuous running. A clean, bright oxide free copper surface resulted, figure 1b. This was adjudged to be a physical, not a chemical phenomenon. The water pH was tested to be neutral and nonane samples analysed by gas chromatography revealed no significant changes from the pure n-nonane initially introduced to the rig. Further, the effect did not occur when condensing the pure fluids. This phenomenon made it possible to carry out condensing tests on the same surfaces in their oxidised and oxide free states. In this context, the oxidised copper surface was obtained from the oxide free one by allowing very small amounts of air into the rig over a period of one week. Normal operation at just over atmospheric pressure was then recommenced. In all cases and at all temperature driving forces, condensation occurred on the oxide free surfaces in the standing drop film mode and in the channelling mode on the oxidised surfaces, figure 1.

Reduction of Data and Estimated Errors.

Condensing surface wall temperature estimation is described above. Liquid/vapour interface temperatures in both azeotropic and non-azeotropic tests were assumed to be azeotropic. Thus $\Delta T_f = T_A - T_w$. Heat fluxes were calculated from measurements of condensate flowrate and composition. Accumulated errors in ΔT_f and q, including errors in measurements, properties and azeotropic temperature were estimated to amount to a standard deviation of ±7% in α_f at ΔT_f=4degC, falling to ±3% at 30degC for the vertical flat plate. Corresponding standard deviations in α_f for the tube configuration tests were ±6% at ΔT_f=4degC and ±3% at 18degC. Repeat measurements made over a period of two weeks indicated a scatter of results of about ±4% in α_f.

5. RESULTS AND DISCUSSION

Azeotropic vapour was condensed in both flat plate and tube configurations. Two cooling side conditions were imposed in the tube column tests, equal wall temperatures in all six tubes or equal cooling water velocities in all six tubes. In each case cooling water flowed in a single pass with the same inlet temperature to each tube. Two nonane rich vapour mixtures, y_{iv} =0.33 and 0.23(mass) steam, compared to 0.39 in the azeotrope, were condensed on the flat plate. Only the y_{iv}= 0.23 nonane rich mixture was used in the tube configuration tests with equal tube wall temperatures. In all the flat plate tests ΔT_f ranged from 2 to 36degC. The corresponding range in the tube column tests was 3 to 23degC.

The experimental results are presented in figures 4-7. In addition predictions of condensate composition in the nonane rich tests, using equations (23-26), are superimposed on figures 6 and 7. Heat transfer coefficients, α_f, predicted by the general channelling flow model, equations (1),(6),(7) or (8) and by the standing drop film model, equations (15-19), are superimposed on figures 4 and 5. Calculations of α_f were based on predicted condensate compositions in the case of the nonane rich tests.

Experimental Results.

In both configurations α_f fell with rise in ΔT_f, when condensing on the oxidised surfaces where channelling flow occurred. On the oxide free surfaces, where the standing drop film pattern existed, α_f is independent of ΔT_f within experimental error. When condensing on the oxidised tubes, with equal cooling water flowrates, α_f was lower, figure 5, than when the wall temperatures were the same. On the oxide free surface the decrease in α_f with ΔT_f is less pronounced and, if anything, figure 5 shows the higher α_f when cooling water flowrates are equal.

The nonane rich experimental data is shown in figures 6 and 7. The effect of mass diffusion resistance to flow of nonane to the condensing surface is evident. Condensing on the plate at y_{iv} =0.33 in standing drop film mode, x_{ic}/y_{iv} varies from 0.61 at ΔT_f=2degC to 1.00 at ΔT_f=33degC. Also on the oxide free surface, at richer nonane inlet conditions, y_{iv} =0.23, x_{ic}/y_{iv} ranged from 0.52 at ΔT_f=2.3degC to 0.97 at 36.7degC. Greater scatter of this set of data occurs, figure 6, partly due to greater scatter than normal in the inlet vapour composition. At y_{iv} =0.23 in the channelling flow mode, mass diffusion resistance was much less than in the standing drop film tests at y_{iv} = 0.22 and even less than at y_{iv} =0.33, much closer to the azeotrope.

At all values of ΔT_f mass diffusion resistance to condensation appears to be greater in the tube column configuration than with the vertical flat plate. However, the values are not nearly so much affected by mode of condensation, figure 7. The authors cannot explain these facts except to note that the channelling flow mode is able to develop on the vertical surface whereas it is continually interrupted by drainage in the tube column. Further, the sweeping effect on the tube surfaces below the top tube, due to drainage from tube to tube in channelling flow, is similar to that which occurs when departing water drops sweep the surface in the standing drop film mode. The pattern of condensation is therefore much the same and there seems no reason to believe that there is any difference between the diffusion resistances of the two modes of condensation. However, in channelling flow on the vertical surface, the liquid phases are well separated. Further, the vapour enters the condenser at the dew point which, at y_{iv} =0.23, is 17-18degC above the azeotropic temperature. It is possible that the interface temperature on the nonane channel lies between T_A and the dew point. Thus the actual mean ΔT_f may be higher than the value ($T_A - T_w$) assumed in plotting figure 6. This would have the effect of moving the channelling flow points in figure 6 to the right, closer to the standing drop film values.

The low mass transfer resistance inferred in figure 6 for channelling flow on the oxidised vertical plate is reflected in high values of α_f, figure 4. Indeed, these values are almost as high as those measured when condensing the azeotrope on the same surface. α_f falls off monotonically as the vapour mixture condensed in standing drop film mode becomes richer in nonane, figure 4. On the tube column, α_f is only slightly higher when condensing the azeotrope on the oxide free surface than the nonane rich mixture on the oxidised surface. The lowest α_f obtained in the tests were for condensing the nonane rich mixture on the oxide free surface.

Theory versus Experiment.

The general channelling flow model predicted α_f well for condensation of the azeotrope on the oxidised vertical plate and on the tube column, with equal tube wall temperatures, figure 4. The shared surface model underestimated α_f by 13% for the vertical plate and by 7-8% for the tube column over the whole range of ΔT_f, table 1. However, allowing for experimental error, the equal cooling water flowrate data lies about half way between the model predictions.

For condensation of the azeotrope on the oxide free vertical surface, the simplified Polley model curve, figure 4, is seen to fall on the data points around 15degC but to overpredict by 8% at ΔT_f=5degC and underpredict by the same amount at 25degC. Poorer agreement is obtained in the case of the tube column. The model, which was developed primarily for vertical surfaces, overpredicts α_f by between 7 and 10%.

Predictions of condensate water content are generally higher than the data for the nonane rich vapour tests on the vertical surface, figure 6. There is reasonable agreement with the data for the standing drop film mode but the predictions are much higher than the data for channelling flow. As discussed above, the interface temperature on the nonane channels may be higher than T_A. This causes uncertainty in the true temperature driving force in the experiments and renders the diffusion model used invalid. Differences between α_f data and predictions for nonane rich vapour condensation, figure 4, are primarily due to errors in predicting x_{ic}. Thus in the standing drop film mode at y_{iv} =0.33 predictions of α_f agree well with the data but underpredict it by up to 17% when y_{iv} =0.22. The corresponding underprediction in the channelling flow mode is 25%.

Despite generally good prediction of condensate composition for the tube column, predictions for the channelling flow mode show a different trend with ΔT_f than the data, figure 5. α_f is underestimated by 23% at ΔT_f=5degC although agreement is good at high ΔT_f. The general model was used in these predictions and even greater underestimates of the data are obtained if the shared surface model is used. This underprediction of non-azeotropic vapour condensation over tubes in the channelling flow mode has been experienced in condensing trichloroethylene/steam mixtures over a tube bundle (Musa et al.,1988). Again it may be a result of underestimating the temperature driving force. Agreement between theory and experiment is better when condensing in the standing drop film mode. The data lies below the predictions by about 10%.

CONCLUSIONS

The following conclusions may be drawn:-

(a) If data suitable for application to long term industrial operation of condensers with 2-phase liquid condensate is to be measured in the laboratory, continuous condensation for long periods is necessary to stabilise the mode of condensation.

(b) The standing drop film mode should be assumed in design of condensers with copper surfaces to operate in low oxygen conditions for long periods.

(c) The channelling flow results obtained for the steam/nonane system are better predicted by the general channelling flow model than by the shared surface model. More work is required to compare the two models under stable conditions for other aqueous systems.

(d) A simplified version of Polley's model has been shown to fit the standing drop film mode of condensation of the steam/nonane azeotrope on a vertical surface and a column of horizontal tubes.

(e) More work is required on condensing non-azeotropic vapours. In particular measurements are required to determine whether non equilibrium interface temperatures occur in channelling flow.

ACKNOWLEDGEMENTS

The authors are grateful to Esso Petroleum, Scotland and Northern Ireland for Scholarship support (H.J.Hoon)

REFERENCES

Bernhardt,S.H., Sheridan,J.J. and Westwater,J.W.,1971, "Condensation of immiscible liquids", AICHE, Symposium Series, Vol.68,Pt.118,pp.21-37.

Bird,R.B., Stewart,W.E. and Lightfoot,E.N.,1960,"Transport Phenomena", Wiley International, New York.

Boyes,A.P. and Ponter,A.B.,1972,"Condensation of immiscible binary systems",CPE-Heat Transfer Survey,pp.26-30.

Deakin,A.W., 1976,"Condensation of vapours of binary immiscible liquids", Ph.D. Thesis,Birmingham University

Glicksman,L.R. and Hunt,A.W., 1972, "Numerical simulation of dropwise condensation", Int.Jl. Heat Mass Transfer,Vol.15,pp.2251-2269.

Graham,C. and Griffith,P.,1973,"Drop size distribution and heat transfer in dropwise condensation", ibid.,Vol.16,pp.337-346.

Hoon,H.J., 1981, "Condensation of binary vapours of immiscible liquids", Ph.D. Thesis, Heriot-Watt University.

Kirkbride,C.C,1934,"Heat transfer by condensing vapour on vertical tubes",Ind.Eng.Chem.,Vol.26,Pt.4,pp. 425-428.

Labuntsov,D.A., 1957, "Heat transfer in film condensation of pure steam on vertical surface and horizontal tubes", Teploenergetika,Vol.4,Pt.7,pp.72-80 (English Translation, CTS454, National Lending Library, Boston Spa, England).

Musa,M.N., Gray,W. and Burnside,B.M., 1988, "Condensation of a binary vapour of immiscible liquids over a tube bundle", Procs.IMechE/IChemE 2nd U.K. Conference on Heat Transfer, Vol.2,pp.987-1001.

Polley,G.T., 1976, "Condensation of binary mixtures of vapours of immiscible liquids", Ph.D. Thesis, Loughborough University of Technology.

Polley,G.T. and Calus,W.F., "The effect of condensate pattern on heat transfer during the condensation of binary mixtures of vapours of immiscible liquids", Procs.6th.Int. Heat Transfer Conf., Vol.2,pp.471-476.

Rose,J.W. and Glicksman,L.R.,1973,"Dropwise condensation -The distribution of drop sizes", Int.Jl.Heat Mass Transfer,Vol.16,pp.411-425.

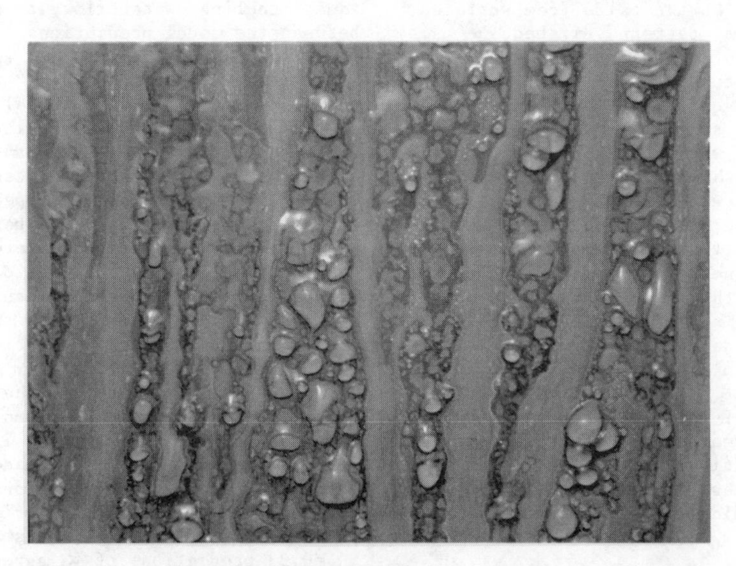

(a) channelling flow

Fig.1 Patterns of condensate flow: steam/nonane system

Fig.1(b) Standing drop film flow (nonane/water system)

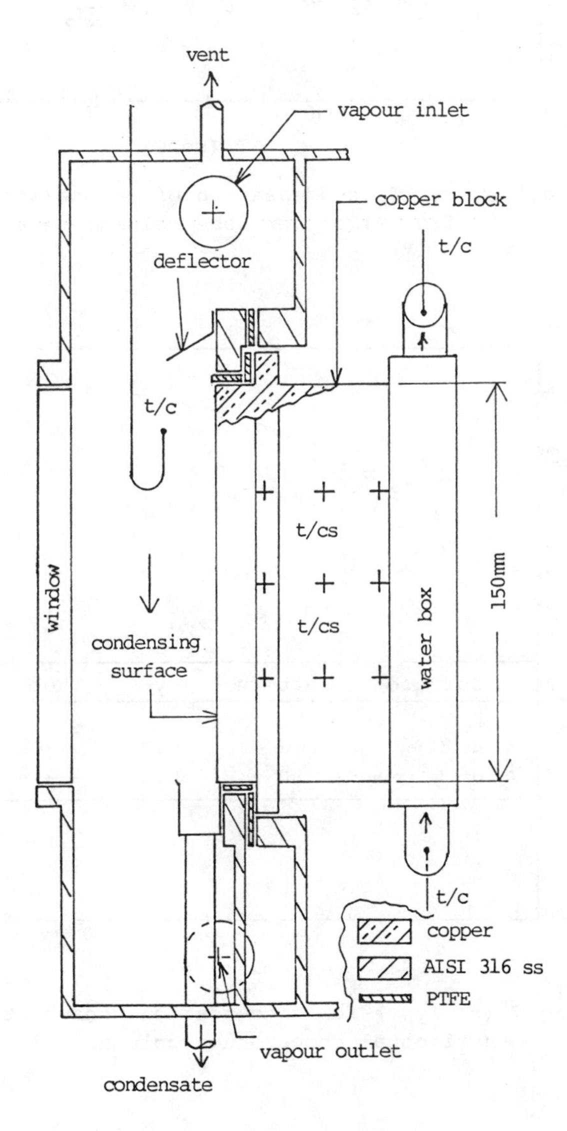

Fig.2 Vertical flat plate condenser

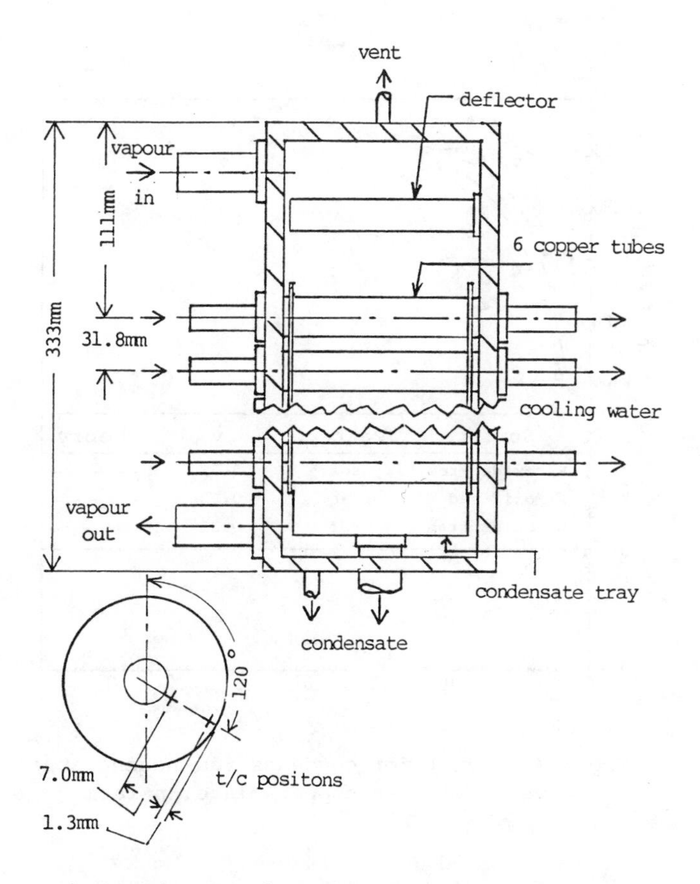

Fig.3 Horizontal six tube single column condenser (only three tubes shown)

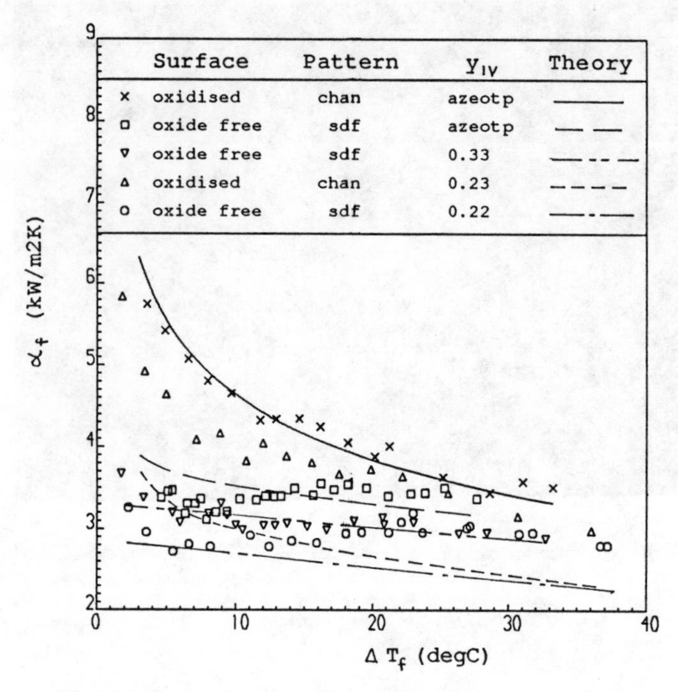

Fig.4 α_f v. ΔT_f :condensation of nonane/steam vertical flat copper plate: p=1atm

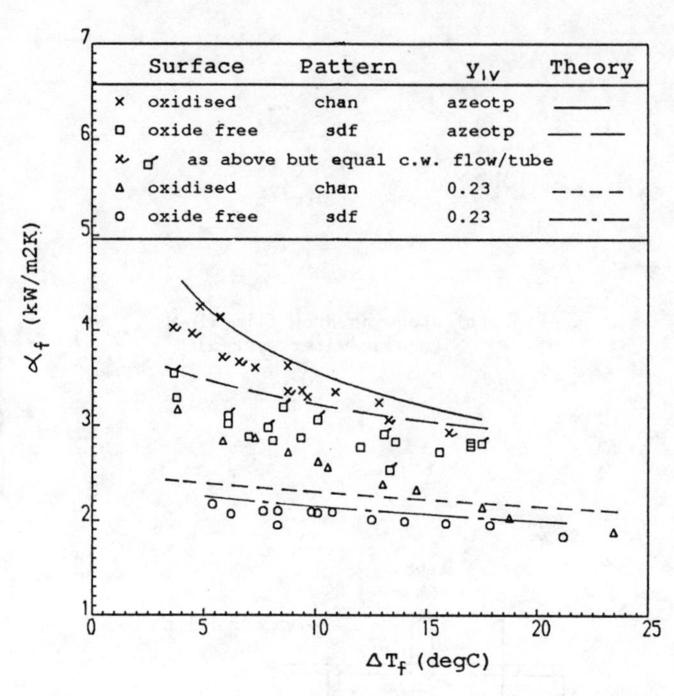

Fig.5 α_f v. ΔT_f :condensation of nonane/steam horizontal copper tube column: p=1atm

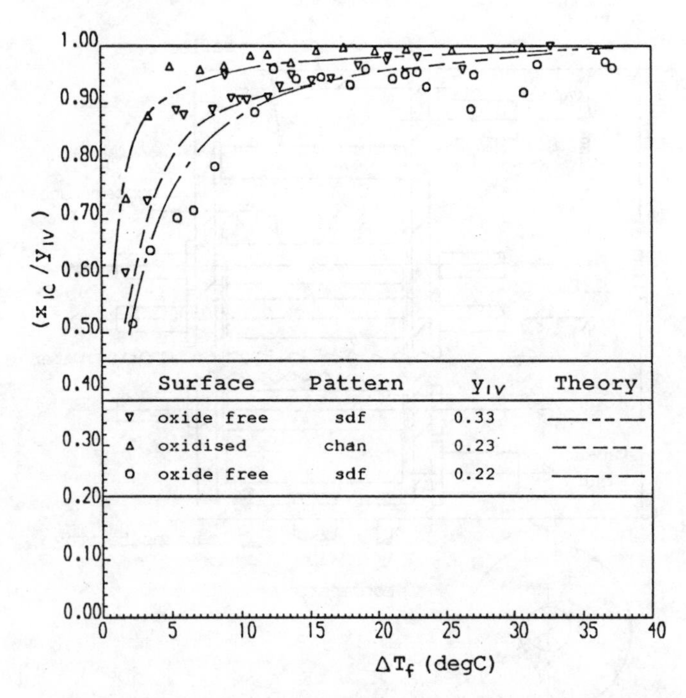

Fig.6 (x_{lc}/y_{lv}) for condensation nonane/steam vertical flat copper plate: p=1atm

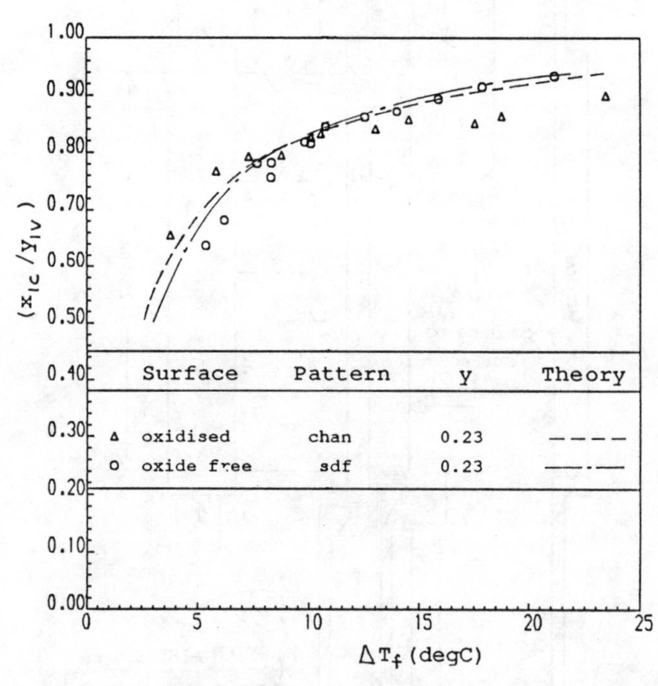

Fig.7 (x_{lc}/y_{lv}) for condensation nonane/steam horizontal copper tube column: p=1atm

CONDENSING HEAT TRANSFER ON A HEMISPHERICAL BODY

A. M. Jacobi
Department of Mechanical Engineering
The Johns Hopkins University
Baltimore, Maryland

V. W. Goldschmidt
School of Mechanical Engineering
Purdue University
West Lafayette, Indiana

M. C. Bublitz
Allied-Signal Aerospace Company
South Bend, Indiana

D. R. Tree
School of Mechanical Engineering
Purdue University
West Lafayette, Indiana

ABSTRACT

The boundary layer equations are applied to obtain a solution for the heat transfer on a condensing hemispherical surface. The method of solution does not restrict itself to assuming a zero condensate thickness at the boundary. It is demonstrated that excluding the radius of curvature effects in the momentum equation leads to an insignificant error. However, the solution is dependent on the boundary condition assumed for the condensate layer. The analysis is generalized for axisymmetric bodies. Comparisons are made to data, but only with limited success.

NOMENCLATURE

c_p	specific heat $(kJ/kg \cdot K)$
D	Characteristic Length (diameter of hemisphere) (m)
Gr_c	condensation Grashof number $= \left(g\frac{(\rho-\rho_v)D^3}{\rho\nu^2}\right)$
g	acceleration due to gravity (m/s^2)
Ja	Jacob number $= \left(\frac{c_p \Delta T}{\lambda}\right)$
k	thermal conductivity $(W/m \cdot K)$
$\dot{m}''$	condensation mass flux $(kg/s \cdot m^2)$
Pr	Prandtl number $= \left(\frac{c_p \mu}{k}\right)$
r	radial coordinate (m)
T	Temperature (K)
u	Streamwise velocity in condensate layer (m/s)
x	Streamwise coordinate (m)

Greek

δ	condensate layer thickness (m)
ϕ	Streamwise angle (rad)
μ	dynamic viscosity of liquid phase $(N \cdot s/m^2)$
ρ	density (kg/m^3)
λ	latent heat of condensation (kJ/kg)
ν	kinematic viscosity of liquid phase (m^2/s)

Subscripts

o	initial, at $\phi = 0$ or $x^* = 0$
sat	at saturation conditions
surf	at the surface
v	of the vapor

Superscripts

*	dimensionless

INTRODUCTION

Laminar filmwise condensation has received a great deal of attention since the pioneering work of Nusselt (1916a,b). The Nusselt-Rohsenow theory accurately predicts heat transfer behavior under many circumstances, and much progress has been made in extending the theory. Sparrow and Gregg (1959a,b) formulated the problem in a boundary layer context, and many subsequent investigators have employed this approach.

The works of Koh, et al. (1961), Chen (1961a,b), Koh (1962), and Shekriladze and Gomelauri (1966) were directed at accounting for shear at the liquid/vapor interface. Dhir and Lienhard (1971) and Lienhard and Dhir (1974) extended the analysis for axisymmetric bodies with nonuniform gravity, and for nonisothermal bodies. Pressure gradient effects were accounted for by Rose (1984), and saturation state and surface tension variation were studied by Jacobi and Goldschmidt (1989). Thorough reviews of the subject are available, e.g. Westwater (1980), Lee and Rose (1982), and Jacobi (1987).

The Dhir and Lienhard (1971) analysis neglects curvature effects, and perscribes a zero condensate layer thickness at the boundary. The purpose now is to explore an axisymmetric geometry in which these effects may be significant. In particular, the case of a hemisphere is to be analyzed. Furthermore, a generalization of the analysis is given which removes the need to perscribe a zero condensate layer thickness at the boundary. These effects are explored, and compared to previous analytical studies and rather limited experimental work.

ANALYSIS

The Hemispherical Case

The model now considered is shown in Fig. 1. The analysis will be developed without specifying an initial condition on the condensate layer thickness, as for this geometry, the boundary condition may not be well known. This is addressed later. The following assumptions are made:

1. Condensation takes place from a quiescent, saturated, pure vapor onto an isothermal, convex-down hemisphere.
2. The condensate film is laminar, has constant thermophysical properties, and its inertial forces are negligible compared to the gravitational and viscous forces. The condensate layer has velocity only in the streamwise ϕ, direction, i.e. $U_r = U_\theta = 0$, and $U_\phi(r, \phi)$.
3. Shear imparted by the vapor on the moving condensate layer is negligible.
4. Heat transfer through the condensate layer is by conduction in the radial direction, and curvature effects are negligible in the conduction equation.

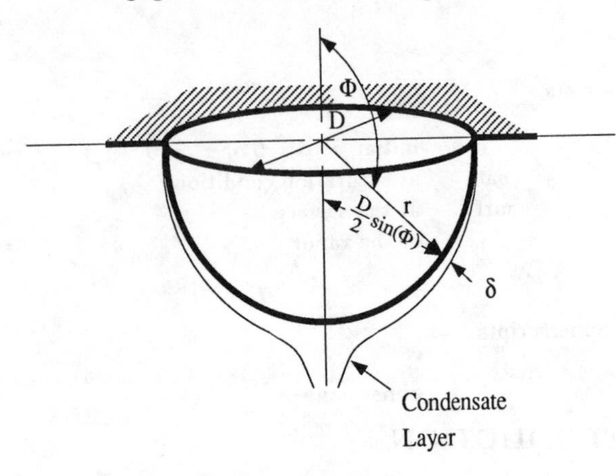

Fig 1. Hemispherical Geometry

5. The temperature is continuous across the liquid/vapor interface, and the saturation state is not influenced by interface curvature.
6. The boundary layer equations are applicable.

With these simplifying assumptions, the conservation of mass, momentum, and energy may be written as:

$$\dot{m}'' = \frac{2\rho}{D\sin\phi}\frac{\partial}{\partial\phi}\int_{\frac{D}{2}}^{\frac{D}{2}+\delta} u\sin\phi\, dr \tag{1}$$

$$\frac{1}{r^2}\frac{\partial}{\partial r}\left(r^2\mu\frac{\partial u}{\partial r}\right) + (\rho - \rho_v)g\sin\phi = 0 \tag{2}$$

and

$$\dot{m}'' = \frac{k\Delta T}{\lambda\delta} \tag{3}$$

where

$$\Delta T = T_{sat} - T_{surf} \tag{4}$$

Integrating the momentum equation and imposing a zero velocity at the hemisphere's surface $\left(r = \frac{D}{2}\right)$, and zero shear at the liquid/vapor interface $\left(r = \frac{D}{2} + \delta\right)$ yields the velocity distribution:

$$u(r,\phi) = \frac{(\rho - \rho_v)g\sin\phi}{6\mu}\left[\frac{2(\frac{D}{2}+\delta)^3}{\frac{D}{2}} + \left(\frac{D}{2}\right)^2 - \frac{2(\frac{D}{2}+\delta)^3}{r} - r^2\right] \tag{5}$$

Following Rose (1984) and Jacobi and Goldschmidt (1989), equations (1) and (3) are combined, and substituting equation (5) for the velocity profile, leads to:

$$\frac{d\delta^*}{d\phi} = \frac{N(\phi,\delta^*)}{D(\phi,\delta^*)} \tag{6}$$

where

$$N(\phi,\delta^*) = 6\frac{Ja}{PrGr_c} - \delta^*\cos\phi\{2\delta^* + 10\delta^{*^2} + \frac{68}{3}\delta^{*^3} + 16\delta^{*^4} - (1+2\delta^*)^3\ln(1+2\delta^*)\}$$

$$D(\phi,\delta^*) = \delta^*\sin\phi\{1 + 10\delta^* + 34\delta^{*2} + 32\delta^{*^3} - 3(1+2\delta^*)^2\ln(1+2\delta^*) - (1+2\delta^*)^2\}$$

and

$\delta^* = \frac{\delta}{D}$; dimensionless condensate layer thickness

$Pr = \frac{c_p\mu}{k}$; Prandtl number

$Ja = \frac{c_p\Delta T}{\lambda}$; Jacob number

$Gr_c = \frac{g(\rho-\rho_v)D^3}{\rho\nu^2}$; a condensation Grashof number

The last grouping is nothing more than a modified Grashof number, where the volumetric thermal expansion coefficient becomes $\frac{1}{\rho}\frac{\rho-\rho_v}{\Delta T}$. Although scaling with $D/2$ may seem more consistent from a boundary layer point of view, it will become apparent that scaling with the characteristic length appearing in the customary Nusselt number makes for very convenient extension and generalization of this analysis.

Eqn. (6), together with the initial conditions, gives a solution for the development of the condensate thickness. For the case of a complete sphere, symmetry would require

$$\left.\frac{d\delta^*}{d\phi}\right|_{\phi=0} = 0 \tag{7}$$

This is formally staisfied (implying that the boundary layer equations are applicable even at the start-up) when

$$2\delta_0^{*^2} + 10\delta_0^{*^3} + \frac{68}{3}\delta_0^{*^4} + 16\delta_0^{*^5} - \delta_0^*(1+2\delta_0^*)^3\ln(1+2\delta_0^*) = 6\frac{Ja}{PrGr_c} \tag{8}$$

In a like manner, any initial condition may be imposed for any section of a spherical condensing surface. This offers a decided advantage over the Dhir and Lienhard (1971) approach, where a zero initial condition was imposed for $\delta_0^* = 0$. As demonstrated later, some geometries, e.g. the one being considered, are sensitive to the initial condition. In further contrast to the Dhir and Lienhard (1971) approach, curvature in the condensate layer was accounted for in the momentum equation. However, as discussed later, this makes very little difference, and the simpler model given by

$$\frac{d\delta^*}{d\phi} = \frac{\frac{Ja}{PrGr_c} - \frac{4}{3}\delta^{*^4}\cos\phi}{2\delta^{*^3}\sin\phi} \tag{9}$$

yields satisfactory results for spherical geometries. Eqn. (9) is the result of carrying out the analysis with no curvature in the momentum equation. If one imposes a zero δ_0^*, it is equivalent to the Dhir and Lienhard (1971) analysis.

In the particular case of a hemisphere, as the local Nusselt number is given by:

$$Nu = \frac{hD}{k} = \frac{1}{\delta^*} \tag{10}$$

an average Nusselt number can be determined (together with the solution of Eqn. (6)) from:

$$\overline{Nu} = \int_{\frac{\pi}{2}}^{\pi} \frac{\sin\phi}{\delta^*} d\phi \tag{11}$$

It is convenient to cast the results into the form:

$$\overline{Nu} = \gamma \left(\frac{Ja}{PrGr_c}\right)^{-\frac{1}{4}} \tag{12}$$

γ results from solution of Eqn. (6) with (11). This may seem somewhat arbitrary, as γ is a function of the boundary condition, however, it is the customary representation, and is very convenient for comparing results.

Eqn. (6) can be integrated with a fourth order Runge-Kutta method, and an average Nusselt number obtained through a fourth order Simpson's rule integration for a particular value of δ_0^*. The value of γ resulting from these integrations for a hemisphere is shown as a function of the initial condition in Fig. 2. The numerical error is estimated to be less than 0.05%.

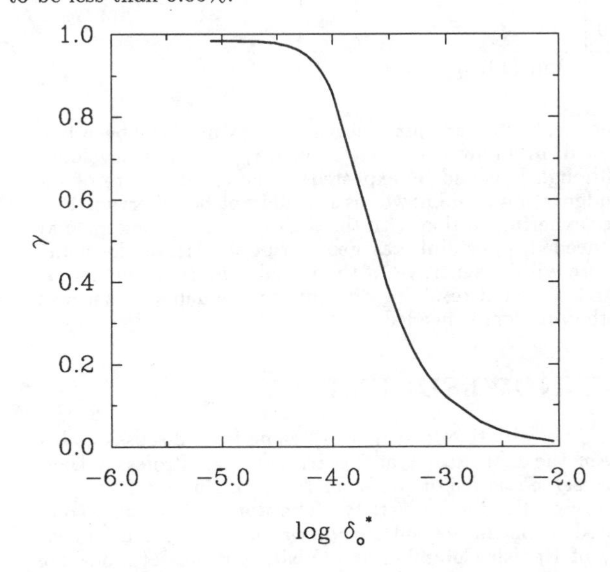

Fig 2. Heat Transfer Dependence on Boundary Condition (Hemisphere)

Discussion of the Hemispherical Case

The results presented in Fig. 2 illustrate the significance of the initial condition in the hemispherical case. Appendix A compares these results to a rather limited (and somewhat restricted) set of data. The data suggest that $\gamma \approx 0.12$, while for the zero-condensate-layer-thickness boundary condition $\gamma \approx 0.99$.

Consideration of the parent equations leads to the recognition that applying the boundary layer equations would require the case of $\delta_0^* = 0$ to be accompanied by an infinite slope; a rather unrealistic situation in view of contact angle considerations. A nonzero δ_0^* may seem somewhat artificial, however it does provide a mechanism for exploring these issues.

Generalization of the Analysis

Considering the foregoing analysis, a generalization and extension of the Dhir and Lienhard (1971) analysis is motivated. As such, consider the geometry shown in Fig. 3. By defining

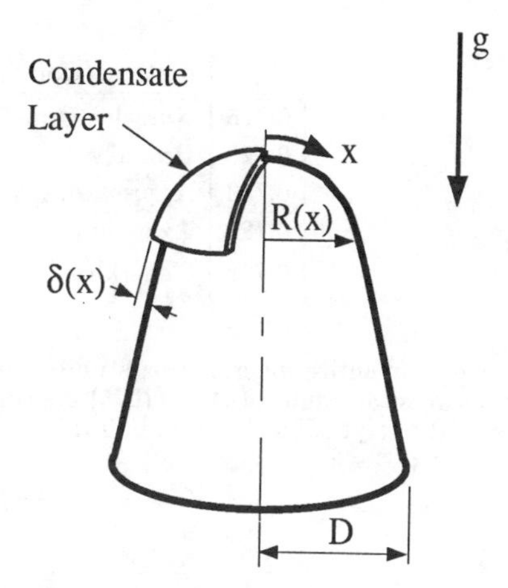

Fig. 3 General Axisymmetric Geometry

$x^* = \frac{x}{D}$	; dimensionless streamwise coordinate.
$R^*(x^*) = \frac{R(x^*)}{D}$	; dimensionless axisymmetric curvature
and $g^*(x^*) = \frac{g(x^*)}{g}$	; dimensionless x^*-component gravitation.

It can be shown that in general;

$$\frac{d\delta^*}{dx^*} = \frac{3\frac{Ja}{PrGr_c} - \delta^{*^4}\left(\frac{dg^*}{dx^*} + \frac{g^*}{R^*}\frac{dR^*}{dx^*}\right)}{3\delta^{*^3}g^*} \tag{13}$$

governs condensate layer growth on any axisymmetric body. A zero boundary condition need not be imposed. In fact, this generalization of Dhir and Lienhard's (1971) work allows any physically meaningful boundary condition to be imposed.

Since the characteristic length, D, was used for scaling, the local Nusselt number may be written as:

$$Nu = \frac{hD}{k} = \frac{kD}{\delta k} = \frac{1}{\delta^*} \tag{14}$$

The average Nusselt number is then:

$$\overline{Nu} = \frac{\int_{x_1^*}^{x_2^*} \frac{R^*(x^*)}{\delta^*} dx^*}{\int_{x_1^*}^{x_2^*} R^*(x^*) dx^*} \tag{15}$$

For the case of a hemisphere, Eqn. (13) reduces to Eqn. (9), and Eqn. (15) to Eqn. (11).

Thus, to estimate the heat transfer behavior during condensation onto any axisymmetric body, Eqn. (13) must be integrated using the relevant boundary condition. Then, Eqns. (14) and (15) provide local and average Nusselt numbers, respectively. The application of this analysis to condensation on a cylinder is compared to previous work in Table 1.

Table 1: Cylindrical Geometry

γ	Method/Comments	Source
0.725	Nusselt estimates his integral	Nusselt (1916b)
0.728	Bromely et al. refine Nusselt's estimate	Bromely et al. (1952)
0.733	Approximate boundary layer solution	Sparrow and Greg (1959)
0.729	Axisymmetric, $\delta_0^* = 0$	Dhir and Lienhard (1971)
0.728	Generalized, $\frac{d\delta_0^*}{d\phi}\Big\vert_{\phi=0} = 0$	Present work

For the case of an entire sphere, fewer studies have been done. The experimental results of Dhir (1975) are compared along with several analytical techniques in Table 2.

Table 2: Spherical Geometry

γ	Method/Comments	Source
0.785	Axisymmetric, $\delta_0^* = 0$	Dhir and Lienhard (1971)
0.803	Boundary layer approximations	Yang (1973)
0.828	Generalized, $\frac{d\delta^*}{d\phi}\Big\vert_{\phi=0} = 0$	Present work
0.828	Integration of Eqn. (6) $\frac{d\delta^*}{d\phi}\Big\vert_{\phi=0} = 0$	Present work
0.785	Quasisteady experimental, $\pm 13\%$	Dhir (1975)

In the cylindrical and spherical geometries, curvature effects in the momentum equation are negligible, always having an influence of less than 0.3% on γ. The data of Dhir (1975) do not allow one to evaluate which method of Table 2 is superior.

CONCLUSIONS AND SUMMARY

The boundary layer equations can be applied to obtain a solution for the hear transfer on a condensing hemisphere, accounting for curvature and the possibility of a nonzero condensate layer thickness at the boundary. The results demonstrate that while the curvature effects are insignificant, the boundary condition can affect the solutions.

The effect of the boundary condition is seen in terms of the γ coefficient in Eqn. (12). Two observations are in order:

1. For a hemisphere, the value of γ is bound between 0 and 1, depending on the boundary condition. (For a sample, yet somewhat inaccurate set of experiments, values were seen in Appendix A to be close to 0.12).
2. The theoretical value of γ depends on the boundary condition, determinable from careful experiments.

The method of solution when generalized for a cylinder and a sphere shows close agreement with other solutions. The hemispherical condensation data due to Bublitz (1988) do not compare favorably with the analysis, if δ_0^* is taken as zero. This may be due to imposing an incorrect $\delta_0^* = 0$ boundary condition, or to a model which completely fails, or to poor data. It is certainly possible that the perscribed boundary condition is inappropriate, and this problem has demonstrated sensitivity to the boundary condition. It is unlikely that the model fails, as it succeeded in all other geometries. The experiments conducted by Bublitz (1988) were not solely directed at the condensation heat transfer coefficients, and adequate degassing may not have been performed, or filmwise condensation may not have developed (although it would be expected). The exact nature of the condensation is unknown, as it could not be observed. Further experimentation with the hemispherical geometry may be needed. Certainly, as geometries similar to the hemisphere will be sensitive to the initial condition, further research aimed at resolving difficulties in modeling this aspect of the problem is needed.

ACKNOWLEDGEMENTS

The initiative for this work came from discussions between the first author and Mark Bublitz. Professor Dave Tree served as major professor to Mark Bublitz during his tenure at Purdue University. The work was in part completed while the second author was hosted by the University of British Columbia as a Visiting Professor; they are also gratefully acknowledged. The authors also acknowledge excellent comments received from reviewers of an earlier manuscript.

REFERENCES

Bromley, L.A., Brodkey, R.S., and Fishman, N., 1952, "Heat Transfer in Condensation, Effect of Temperature Variation Around a Horizontal Condenser Tube", *Engineering and Process Development*, Vol. 44, pp. 2962-2967.

Bublitz, M.C., 1988, "Modeling and Experimental Verification of Vapor Condensing Commercial Cooking Equipment", M.S. Thesis, Purdue University.

Chen, M.M., 1961a, "An Analytical Study of Laminar Film Condensation: Part I - Flat Plates", *Journal of Heat Transfer*, Vol. 83, pp. 48-54.

Chen, M.M., 1961b, "An Analytical Study of Laminar Film Condensation: Part II - Single and Multiple Horizontal Tubes", *Journal of Heat Transfer*, Vol. 83, pp. 55-60.

Dhir, V.K., 1975, "Quasi-Steady Laminar Film Condensation of Steam on Copper Spheres", *Journal of Heat Transfer*, Vol. 97, pp. 347-351.

Dhir, V.K., and Lienhard, J.H., 1971, "Laminar Film Condensation on Plane and Axisymmetric Bodies in Nonuniform Gravity", *Journal of Heat Transfer*, Vol. 93, pp. 97-100.

Jacobi, A.M., 1989, "High Efficiency Boilers: Condensation and Transient Behavior", Ph.D. Dissertation, Purdue University.

Jacobi, A.M., 1987, "Laminar Film Condensation: A Review of the Literature", Ray W. Herrick Laboratories, Report HL87-47, Purdue University, West Lafayette, IN.

Jacobi, A.M., and Goldschmidt, V.W., 1989, "The Effect of Surface Tension Variation on Filmwise Condensation and Heat Transfer on a Cylinder in Cross Flow", *International Journal of Heat and Mass Transfer*, Vol. 32, pp. 1483-1490.

Koh, J.C.Y., 1962, "Film Condensation in a Forced-Convection Boundary-Layer Flow", *International Journal of Heat and Mass Transfer*, Vol. 5, pp. 941-954.

Koh, J.C.Y., Sparrow, E.M., and Hartnett, J.P., 1961, "The Two Phase Boundary Layer in Laminar Film Condensation", *International Journal of Heat and Mass Transfer*, Vol. 4, pp. 69-82.

Lee, W.C., and Rose, J.W., 1982, "Film Condensation on a Horizontal Tube: Effect of Vapor Velocity", Proc. Seventh Int. Heat Transfer Conf., Munich, pp. 101-106.

Lienhard, J.H., and Dhir, V.K., 1974, "Laminar Film Condensation on Nonisothermal and Arbitrary-Heat-Flux Surfaces, and on Fins", *Journal of Heat Transfer*, Vol. 96, pp. 197-203.

Nusselt, W., 1916a, "Die Oberflachenkondensation des Wasserdampfres", *Zeitschriff des Vereines Deutscher Ingenieure*, Vol. 60, pp. 541-546.

Nusselt, W., 1916b, "Die Oberflachenkondensation des Wasserdampfres", *Zeitschriff des Vereines Deutscher Ingenieure*, Vol. 60, pp. 529-575.

Rose, J.W., 1984, "Effect of Pressure Gradient in Forced Convection Film Condensation on a Horizontal Tube", *International Journal of Heat and Mass Transfer*, Vol. 27, pp. 39-47.

Shekriladze, I.G., and Gomelauri, V.I., 1966, "Theoretical Study of Laminar Film Condensation of Flowing Vapour", *International Journal of Heat and Mass Transfer*, Vol. 9, pp. 581-591.

Sparrow, E.M., and Gregg, J.L., 1959a, "A Boundary-Layer Treatment of Laminar Film Condensation", *Journal of Heat Transfer*, Vol. 81, pp. 13-18.

Sparrow, E.M., and Gregg, J.L., 1959b, "Laminar Condensation Heat Transfer on a Horizontal Cylinder", *Journal of Heat Transfer*, Vol. 81, pp. 291-296.

Westwater, J.W., 1980, "Condensation, Heat Transfer in Energy Problems", Japan-U.S. Joint Seminar, Tokyo, pp. 67-77.

Yang, J.W., 1973, "Laminar Film Condensation on a Sphere", *Journal of Heat Transfer*, Vol. 95, pp. 174-178.

APPENDIX A - COMPARISON TO DATA

Partial data of heat transfer on a condensing hemisphere were obtained by Bublitz (1988). The set-up, Fig. 4, consisted of a sending unit (boiler) and a receiving unit (condenser). A vapor, generated in the sender, filled the cavity surrounding the outside surface of the bowl. This hemispherical bowl had a diameter of 0.661 m (2.17 ft), and was constructed of $2.29(10^{-3})$m $(7.5(10^{-3})$ft) thick stainless steel. Controlling the saturation pressure afforded cooking temperature control.

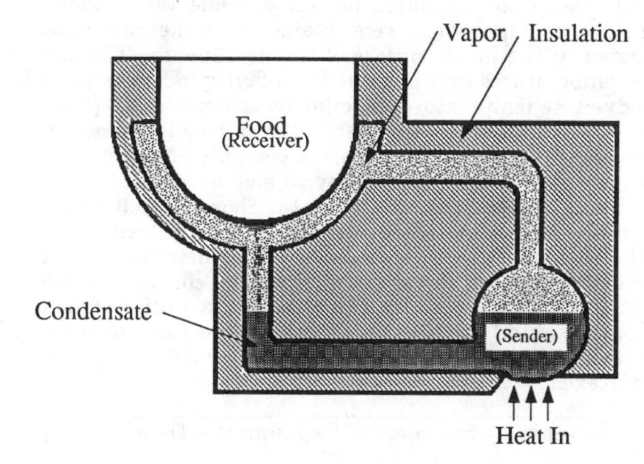

Fig 4. Schematic of Cooking Apparatus

Two different sending units were studied by (Bublitz, 1988), an electric resistance heating scheme, and a gas-fired boiler. No special precautions were taken for surface purity, nor for complete degassing of the distilled water prior to testing, as the attempt was to simulate an actual commercial cooker.

Steady and transient data were recorded. The steady data were obtained by cooling the contents in the receiver at a steady rate and testing after all temperatures reached a steady state value. Under these steady state conditions, the cooling required (and equated to the heat transfer) was determined from flow rates and outlet and inlet temperatures measured for water flowing through a coil immersed on the contents of the bowl. Flow rates were measured with a weigh tank and stop watch. The temperatures were measured with thermocouples squeezed between an outer sleeve and the copper tubing at the coil inlet and outlet. (It is estimated that heat transfer may have had an uncertainty of 20 to 30%.)

Sample data for a transient run are shown in Fig. 5 (Bublitz, 1988). The heat transfer in this case was determined from the temperature rise of the contents of the receiver. (For these cases the uncertainty in the heat transfer obtained from the measurements is estimated to be within 10%.)

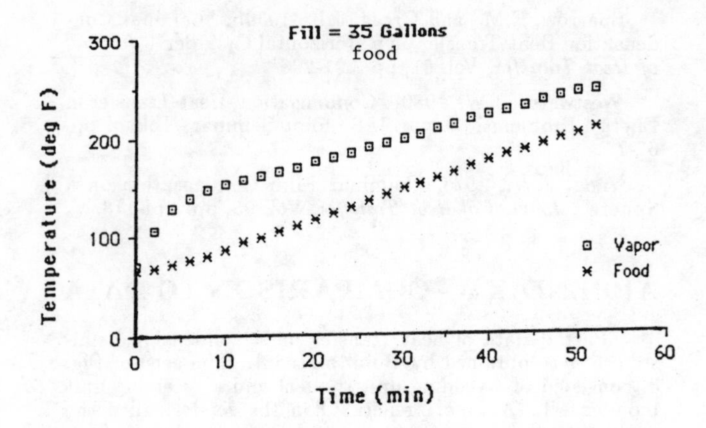

Fig 5. Example Data

The vapor temperatures were estimated as the saturation values for measured pressures, while the condensing surface temperatures were obtained from thermocouples soldered to the inside surface of the hemisphere. The surface temperatures were measured at different locations (with steady-state temperatures differing by as much as 3C (5F)) and were averaged to characterize the surface temperature. Based on these approximations, the computed Nusselt number may have had an uncertainty as high as 50%.

Table 3 summarizes the test data. Shown as well are the corresponding values of Ja, Pr, Gr_c and γ (defined in Eq. (12)). It must be underscored that the condensing medium was distilled water, it was not dionized, no effort was made to boil off noncondensibles, and the exact condition (finish, cleanliness, etc.) of the condensing surface was unknown. No observations to verify the mode of condensation were undertaken.

Table 3 - Summary of Experimental Data

Run #	Tvap K	ΔT K	Pr	Ja	Gr_c $\times 10^{-12}$	$\overline{Nu}$	γ
5s1	320	9.7	3.77	.017	8.33	788	0.12
6s2	330	8.3	3.15	.018	11.48	1590	0.33
7s2	349	10.5	2.30	.019	20.00	903	0.12
trans*	380	26.4	1.61	.050	38.08	802	0.14

* Averaged between 30 and 50 minutes after start-up.

Inspite of the lack of accuracy in the test procedures, except for one case, the value of γ is somewhat consistent, and equal to approximately 0.12. A comparison with the analysis would imply an initial boundary layer thickness of $\delta_0^* \approx 0.001$, which for this particular case corresponds to about 0.7mm (.03 in), which appears reasonable.

Dhir (1975) presents data for quasi-steady laminar film condensation of steam on copper spheres. Dhir's experiments were conducted by suddenly immersing isothermal copper spheres into a nearly quiescent jar of steam. The steam was obtained by boiling water in the jar for at least 1.5 hours. The transient temperatures of the sphere's center were recorded, and together with the saturation temperature these data were interpreted in a quasi-steady fashion to yield average Nusselt numbers for condensation on a sphere. As noted in Jacobi (1989), incipient condensation of water on copper appears to be inherently dropwise. Dhir conducted the experiments with spheres having diameters of 0.0317, 0.0254, and 0.0190m. The spheres were always assumed to be isothermal.

The experimental results due to Dhir (1975) were compared to the analysis in Table 2.

AN EXPERIMENTAL STUDY OF LAMINAR FILM CONDENSATION WITH STEFAN NUMBER GREATER THAN UNITY

R. L. Mahajan
AT&T-Bell Laboratories
Princeton, New Jersey

T. Y. Chu
Sandia Laboratories
Albuquerque, New Mexico

D. A. Dickinson
AT&T-Bell Laboratories
Princeton, New Jersey

ABSTRACT

Experimental laminar condensation heat transfer data is reported for fluids with Stefan number up to 3.5. The fluid used is a member of a family of fluorinated fluids developed in the last decade which have been extensively used in the electronics industry for soldering, cooling, and testing applications. Experiments were performed by suddenly immersing cold copper spheres in the saturated vapor of this fluid, and heat transfer rates were calculated using the quasi-steady temperature response of the spheres. In these experiments, the difference between saturation and wall temperature varied from 0.5 ℃ to 190 ℃. Over this range of temperature difference, the condensate properties vary significantly. For example, viscosity of the condensate varies by a factor of over 50. Corrections for the temperature dependent properties of the condensate therefore were incorporated in calculating the Nusselt number based on the average heat transfer coefficient. The results are discussed in light of past experimental data and theory for Stefan number less than 1. To the knowledge of the authors, this is the first reported study of condensation heat transfer for Stefan number greater than unity.

NOMENCLATURE

Bi	Biot number, $hD/2k_c$
c_p	specific heat at constant pressure
c_{pc}	specific heat of condensing surface
D	sphere diameter
g	gravitational acceleration
h	average heat transfer coefficient
h_{fg}	latent heat of vaporization
h'_{fg}	$h_{fg}(1 + 0.68\ S)$
k	thermal conductivity of condensate
k_c	thermal conductivity of copper
m_c	mass of condensing surface
Nu	average Nusselt number based on sphere diameter D
Nu_x	local Nusselt number, hx/k
Pr	condensate Prandtl number, $\mu c_p/k$)
S	Stefan or Jakob number, $c_p \Delta T/h_{fg}$
T	temperature
t	time
x	distance from the upper stagnation point for a sphere or from the leading edge for a plate

Greek Symbols

ΔT	difference between saturation temperature and sphere surface temperature
ΔT_o	difference between saturation temperature and constant surface temperature
δ	quasi-steady condensate film thickness
δ_{ss}	steady state condensate film thickness
μ	condensate dynamic viscosity
ν	condensate kinematic viscosity
ρ	density
τ	time

Subscripts

$calc$	calculated theoretical value
cp	constant property
exp	experimental
f	condensate
g	vapor
m	mean
ref	reference
sat	saturated
vp	variable property
w	sphere surface

1. Introduction

The classical analysis of laminar film condensation on vertical or inclined surfaces is due to Nusselt (1916). In that analysis the condensate film was assumed to be thin, and convective and inertia effects were considered to be negligible. Within the condensate film, gravity was balanced simply by the viscous force, and the temperature profile in the film condensate was assumed to be linear. From the analysis, the local heat transfer rate, Nu_x, was calculated to be

$$Nu_x = \left[\frac{(\rho_f - \rho_g)\, g\, h_{fg}\, x^3}{4\, \nu\, k\, (T_{sat} - T_w)}\right]^{1/4} . \tag{1}$$

Bromley (1952) performed an analysis using a non-linear temperature profile.

Rohsenow (1956) expanded the calculation to include the effect of liquid cross-flow within the film. The analysis, based on a control volume approach, resulted in a differential-integral equation which was solved by successive approximation. It was shown that the effect of the non-linear temperature distribution in the film can be accounted for by replacing h_{fg} in equation (1) by h^+_{fg}, where

$$h^+_{fg} = h_{fg}(1+\frac{3}{8}S)\ (1 - .1\ S - .0328S^2)/(1 - .1S)^2 \ . \tag{2}$$

For the range $0 < S < 1$, it was shown that equation (2) is very closely approximated by

$$h'_{fg} = h_{fg}\ (1 + .68S) \ . \tag{3}$$

The complete boundary solution, including the inertia and convection effects, was obtained by Sparrow and Gregg (1959). It was shown that except for very low Pr, Rohsenow's approximate analysis was quite adequate.

Koh et al. (1961) applied a boundary layer treatment to include the interfacial shear while Chen (1961) considered analytically the effect of thermal convection, inertia forces and interfacial shear. The Nusselt-Rohsenow approach was extended by Dhir and Lienhard (1971) to include axisymmetric bodies in non-uniform gravity. For a sphere of diameter D, the Nusselt number based on an average heat transfer coefficient was calculated to be

$$Nu = 0.785\ \left[\frac{(\rho_f - \rho_g)\, g\, h'_{fg}\, D^3}{\nu\, k\, (T_{sat} - T_w)}\right]^{1/4} . \tag{4}$$

For a review of these and other earlier studies, see Merte (1973).

A study pertinent to the present investigation is an experimental investigation of quasi-steady laminar film condensation of steam on copper spheres by Dhir (1975). The average heat transfer data was obtained in the range of Stefan number 0.009 - 0.12 and was shown to be within ±13% of the steady state theoretical results of Dhir and Lienhard (1971) and Yang (1973).

All the studies of condensation heat transfer up to the present have been for relatively small sensible heat effects. The contribution of sensible heat is accounted for by correcting the latent heat of vaporization by a sensible heat term in the manner of equation (3). In these studies the correction is considered accurate for Stefan numbers up to 1. The recent study of Sadasivan and Lienhard (1987) showed that the correction factor is weakly dependent on *Pr*. The conventional view expressed in the literature is that this correction is adequate since *S* beyond unity exceeds the range of practical interest (Sadasivan and Lienhard, 1987).

In the last decade, new families of fluorinated fluids having Stefan numbers much greater than 1 in heat transfer applications have become widely used in various industries. Typical applications include electronic cooling and a mass soldering process, "condensation soldering." The process was invented as a method of supplying heat for soldering connector pins to printed circuit boards for telephone switching systems (Chu et al. 1974, Pfahl et al. 1975, Wenger and Mahajan 1979). In this method, articles to be soldered, having pre-deposited solder at joints, are immersed in a body of saturated vapor of a flourinated liquid with a typical saturation temperature of 215 ℃. The basic soldering machine consists of a vessel where vapor of the flourinated liquid is generated continuously by boiling. The vapor is condensed on a condensing coil near the top opening of the vessel. Because the vapor is much heavier than air, a stable body of saturated vapor can be maintained in the vessel between the boiling fluid layer and the condensing coil. The heat transfer to articles immersed in the vapor is rapid and uniform, with absolute control of the maximum temperature. Today, condensation soldering is used throughout the electronics industry.

During condensation heating of articles to be soldered, Stefan numbers as high as 3.5 are encountered. Therefore, it is of theoretical as well as practical interest to understand the effect of large sensible heat in condensation heat transfer. The present paper reports the results of a series of experiments carried out to observe film condensation for Stefan numbers up to 3.5. Condensation in this experiment is quasi-steady, and heat transfer rate is readily calculated by monitoring the temperature response of copper spheres suddenly immersed in saturated vapor of the fluorinated liquid.

2. Fluid Properties

The condensing fluid used in the present investigation is perfluorotriamylamine, $(C_5F_{11})_3N$, a member of the family of the perfluorinated inert liquids manufactured by 3M Co. The fluid is sold as *Fluorinert*^{R} liquid and is commercially identified as FC-70. The physical properties of this fluid at 25 ℃, taken from the 3M Fluorinert Electronic Liquid Product Manual, are given in Table 1.

Table 1. Physical Properties of $FC-70^R$

Typical Boiling Point ℃	215
Pour point ℃	-25
Average molecular weight	820
Surface tension, *dynes/cm*	18
Critical temperature, ℃	335 *
Critical pressure, *atmospheres*	10.2 *
Vapor pressure, *torr*	< 0.1
Solubility in water, *ppm*	8
Solubility of air, *ml/100ml*	22
Density, *g/ml*	1.93
Viscosity, *centistokes*	14.0
Specific heat, *cal/g*−℃	0.25
Heat of vaporization, *cal/g*	16
Thermal conductivity, mW/cm^2−(℃/*cm*)	0.69 *
Coefficient of expansion, *ml/ml*−℃	0.0010

* Estimated
R Trademark

The properties of this fluid that are of significance in determination of the film condensation heat transfer rate are the kinematic viscosity, thermal conductivity, specific heat, density, and heat of vaporization. Some of these are strongly dependent on temperature. For example, data provided by 3M Co. on measurements of viscosity (Table 2) indicate that for temperature over the range from room temperature to the boiling point of the liquid, the dynamic viscosity varies by as much as a factor of approximately 54. A least squares fit of the measured data resulted in the correlation of kinematic viscosity as a function of temperature,

$$\log \log (\nu + .90825) = (14.1226 - 5.69327 \log T) \qquad (5)$$

where *T* is the temperature (*K*) and ν is dynamic viscosity (*centistokes*). The general form of this correlation is that recommended by ASTM Standard D 341-77, "Viscosity-Temperature Charts for Liquid Petroleum Products". The equation fitted the data in Table 2 to an accuracy of ±3%.

Table 2. Liquid Viscosity as a Function of Temperature

Temperature ℃	Measured Value of Viscosity, Centistokes
24	11.88
30	8.61
40	5.87
60	2.88
80	1.65
100	1.09
110	0.91
120	0.76
130	0.66
140	0.57
150	0.50
173	0.38
216	0.25

The measurements of *k*, ρ, and c_p as a function of temperature were obtained at AT&T's Thermal Engineering Laboratory at Bell Laboratories, Princeton. The best fit correlations describing the temperature dependencies of these properties are

$$k = 0.7242718 - 6.353033 \times 10^{-4}\, T \qquad (6)$$

$$c_p = .2468938 + 1.51426 \times 10^{-4}\, T + 1.285630 \times 10^{-7}\, T^2 \qquad (7)$$

$$\rho_f = 1.970381 - 1.829488 \times 10^{-3}\, T \qquad (8)$$

where *k*, c_p, ρ_f, and *T* are thermal conductivity in *mW/cm* − ℃, specific heat in *cal/g* − ℃, density in *g/ml*, and temperature in ℃, respectively. The correlations fit the experimental data to an accuracy of ±2%

3. Experimental Apparatus and Procedure

Condensation experiments were conducted using copper spheres with diameters of 25.4 *mm* and 50.8 *mm* as the condensing surface. A diagram of the experimental set-up and the sphere assembly details are shown in Figures 1 and 2 respectively. The body of the sphere was machined from copper and finished with a smooth surface. It was supported from the bottom on a stainless steel tube to prevent condensate on the tube surface from running onto the sphere, and a ceramic insulator was used to separate the two pieces. A thermocouple used to measure the sphere temperature was positioned at the sphere center, with the leads running through the ceramic and through the stainless steel tube. Epoxy was used to support the thermocouple bead, which was held in place against the copper by the weight of the sphere. Epoxy was also used as a seal where the ceramic entered both the sphere and the steel tube.

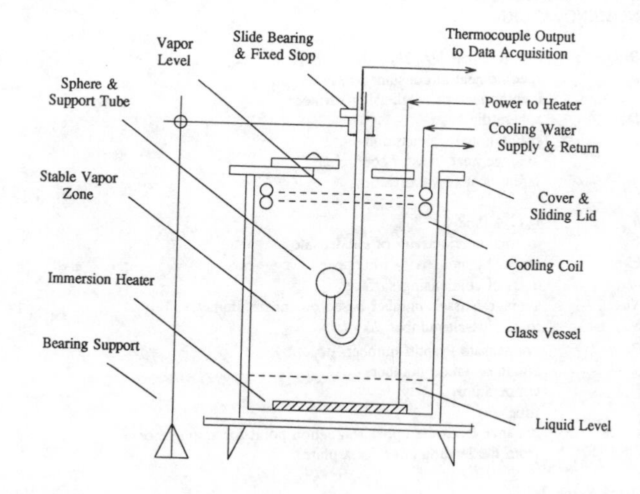

FIGURE 1. Experimental Apparatus

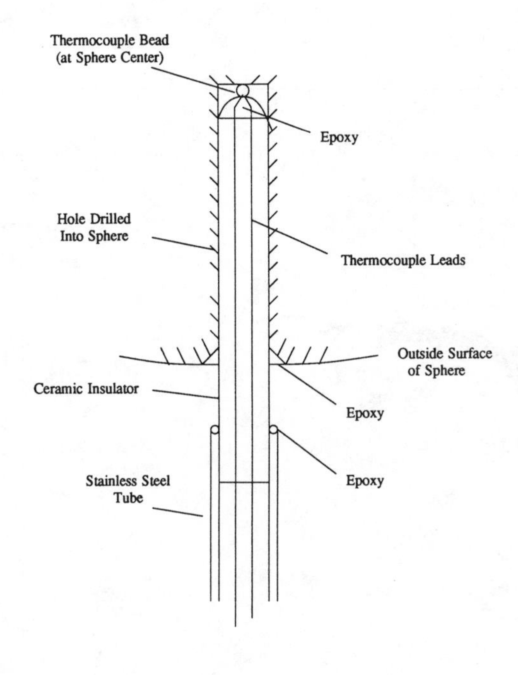

FIGURE 2. Details of Sphere Assembly

The vessel used to contain the sphere and condensing vapor was a cylindrical glass jar with height 46 *cm* and diameter 30 *cm*, as shown in Figure 1. Electric resistance immersion heaters were used to boil the liquid in the base of the jar and water cooled coils at the top were used to continuously condense the vapor, so that a region of saturated vapor was set up when the apparatus was fully heated and operating at steady state. Air filled the space above the saturated vapor zone. Power input to the heaters was adjustable using a variac. The jar was fitted with a stainless steel cover, with a smaller removable cover to allow the sphere to be inserted into the vapor. Hence the entire system was at atmospheric pressure. An equipment stand fitted with a slide bearing was used to hold the stainless steel tube supporting the sphere, with a stop clamped to the tube to fix the vertical position of the sphere within the vapor volume. This arrangement allowed manual insertion of the sphere to be fast, smooth, and repeatable. At the beginning of each run a thermocouple was used to read the actual vapor temperature T_{sat} to be used in the data reduction.

To begin an experimental run, steady state was first established in the jar. The heater power was adjusted so that the vapor/air interface was at the top turn of the condensing coils. This allowed the insertion of the sphere and resulting condensation to have a minimal effect on the vapor height, thus minimizing turbulence at the vapor/air interface and any intermixing of air into the vapor zone. Consistent with the past observations of Chu et al. (1975), the interface was observed to be well defined and stable. Temperature measurements were started just as the sphere, normally at room temperature, was lowered through the cover into position and continued until the sphere temperature was within a few degrees of T_{sat}. A data acquisition system was used to read the sphere temperature at pre-programmed time intervals, selected as 0.5 seconds for both sphere sizes. This interval was long enough to allow negligible error due to variation in the actual time interval used by the acquisition system as well as short enough to allow accurate determination of dT_w/dt at a given measured temperature T_w.

Profiles for T_w and dT_w/dt at the start of a run are shown in Figure 3. A transient period occurs as the sphere enters the vapor and the condensate film is established, followed by the period of quasi-steady condensation, during which a continuous condensate film was observed. As the run proceeds the temperature asymptotically approaches T_{sat}.

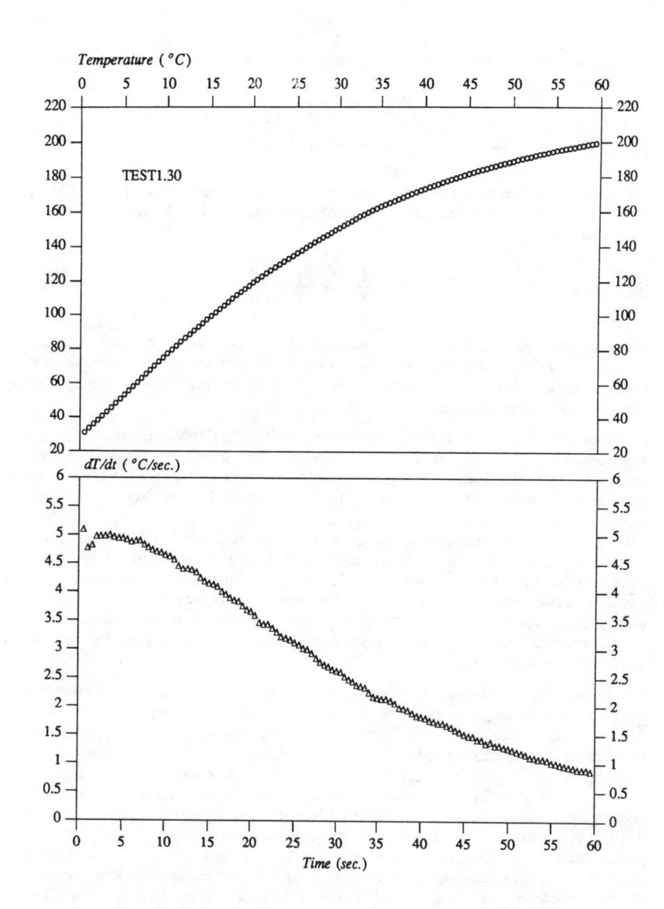

FIGURE 3. (a) Temperature vs. Time, (b) Rate of Temperature Rise vs. Time

4. Data Reduction and Analysis

4.1 Experimental Nusselt Number

A Fortran program was written to process the experimental data, which consisted simply of the temperature measurements at equal time intervals. The heat transfer coefficient h, at a given measured wall temperature, T_w, was determined using the lumped capacity relation,

$$m_c \, c_{pc} \, (dT_w/dt) = hA \, (T_{sat} - T_w) \quad . \tag{9}$$

The experimental Nusselt number Nu_{exp} is then

$$Nu_{exp} = hD/k \quad . \tag{10}$$

The error introduced using equation (9) is negligible if Bi is less than 0.4 (Dhir, 1975). For all the data to be presented Bi is less than 0.05, so the temperature of the sphere can be assumed uniform.

The rate of temperature rise, $dT_w/dt = \Delta T_w/\Delta t$ was calculated using at least two time intervals (1 second) to ensure sufficient accuracy for ΔT_w. At the beginning of a run the rate of temperature rise is about 5 ℃/s with the 25.4 *mm* sphere. The accuracy of the temperature measurement is ±0.1 ℃, so the accuracy of the calculation for h and Nu_{exp} is approximately ±5%. At higher sphere temperatures, with slower heating rate, additional time intervals are added to ensure that ΔT_w is at least 2 ℃. The wall temperature was taken to be the average of the initial and final temperatures. This was used rather than an actual intermediate measurement of temperature to allow the use of odd numbers of intervals. The difference between the average and actual intermediate temperature was typically less than 0.1 ℃.

4.2 Calculated Nusselt Number

In this experimental investigation, the film Reynolds number did not exceed 3, so that laminar film condensation analysis is applicable. For calculation of the theoretical Nusselt number, it must be determined whether the experiment is truly quasi-steady. Furthermore, a relationship between the quasi-steady and steady state calulations needs to be obtained. Dhir (1975) determined the relative liquid film thicknesses for the quasi-steady and steady state conditions and estimated the agreement that could be expected between the measured heat transfer coefficients for the two cases. For a vertical flat plate, local steady state film thickness is

$$\delta_{ss} = \left[\frac{4 \nu k \Delta T_0 x}{g(\rho_f - \rho_g) h'_{fg}} \right]^{1/4} \tag{11}$$

It was shown by Dhir that the governing equation for the quasi-steady film thickness δ can be written as

$$\left[\frac{g(\rho_f - \rho_g)h_{fg}'}{\nu k \Delta T(t)}\right]\delta^3 \frac{\partial \delta}{\partial x} = 1 + \left\{\left[\frac{dT_w}{dt}\frac{1}{\Delta T(t)}\frac{\delta^2}{\alpha}\right] \times \left[\frac{5}{8} + \frac{h_{fg}}{4c_p \Delta T(t)}\right]\right\}. \quad (12)$$

For a sphere, the term g can be replaced by an effective gravitation as derived by Dhir. If the entire right hand side is treated as a constant and the left hand side is integrated, then

$$(\delta/\delta_{ss})^4 = 1 + \left\{\left[\frac{dT_w}{dt}\frac{1}{\Delta T(t)}\frac{\delta^2}{\alpha}\right] \times \left[\frac{5}{8} + \frac{h_{fg}}{4c_p \Delta T(t)}\right]\right\} \quad (13)$$

where δ_{ss} has been substituted using equation (11). The term in the curly brackets represents a ratio of dimensionless thermal diffusion constant and the sensible heat capacity of the film (Dhir, 1975). When this ratio is small, quasi- steady film thickness approaches the steady state value.

It was shown by Dhir that for steam condensing on a copper sphere the ratio δ/δ_{ss} is approximately 1.06 with the minimum value of S = 0.009, so that quasi-steady condensation should result in measured values of heat transfer coefficient within a few percent of those that would be obtained for steady state. For the present case with FC-70 condensing on a 25.4 mm sphere, with $(dT_w/dt)(1/\Delta T)$ approximately 0.035 s^{-1} (from the measured data), an intermediate value of S = 1.5, approximate film thickness of $1.0 \times 10^{-4} m$ (from equation 11), and thermal diffusivity of the condensate of $3.3 \times 10^{-8} m^2/s$, we have $\delta/\delta_{ss} = 1.008$. The agreement is considerably closer than that with steam, due partly to the slower heat transfer with this fluid, but mainly due to the much larger value of S. Clearly the steady state calculations can be used to accurately predict the quasi-steady condensation heat transfer rates for the fluid used in this investigation.

For determining the theoretical values of Nusselt number, Nu_{calc}, the full boundary layer solutions of Sparrow and Gregg (1959) can be used. However, as noted there, for $Pr > 10$, the full boundary layer solutions are very close to the approximate calculations of Rohsenow (1956). For FC-70 condensate, Pr varies from 320 at room temperature (25 ℃) to approximately 8 at the saturation temperature (215 ℃). Equations (2) and (4) are therefore appropriate for obtaining theoretical estimates of Nu. Also note that although equation (3) is prescribed in Rohsenow's paper for use in the range $0 < S < 1$, our calculations of h using both equations (2) and (3) for values of S much greater than 1 (we performed calculations for S up to 20) indicate that difference between the two calculations is at the most 0.5%. In the results to follow, we have therefore chosen to use the simplified equation (3) in conjunction with equation (4) to calculate Nu_{calc}.

The experimental heat transfer data will be presented relative to the theoretical values calculated in the manner above in terms of the ratio Nu_{exp}/Nu_{calc}. The wall temperature will be presented in terms of the Stefan number, S.

5. Results and Discussion

Results for three runs using the smaller sphere are shown in Figure 4. The temperature T_{ref} used for calculation of fluid properties was the average of T_w and T_{sat}. Attention will be focused on TEST1.30, which is a typical run with the sphere initially at room temperature, corresponding to an initial Stefan number of nearly 3.5.

The sequence of points measured during the run moves to the left as the sphere heats up. It appears that the Nusselt number ratio is fairly constant for Stefan numbers between approximately 0.3 and 2.3 and drops off at both ends of this range. There are a number of non-ideal effects that contribute to this behavior, including disturbance to the vapor zone due to insertion of the test object, the effect of property variation, and effect of non-condensable gases. Each of these effects will be addressed in the following sections.

5.1 Effect of Thermal Transients and Initial Disturbance

The time required to achieve steady state after suddenly dropping the temperature of a sphere already immersed in saturated vapor was obtained by Sparrow and Siegel (1959) as

$$t_{ss} = \left[\frac{h_{fg}\rho_f \mu x}{g(\rho_f - \rho_g)k\Delta T_0}\right]^{1/2} \times \left[1 + \frac{c_p \Delta T_0}{2h_{fg}}\right] \quad (11)$$

For the present case with FC-70, t_{ss} is approximately 0.25 to 0.50 seconds. However, this accounts for the thermal transient only and does not include the effect of lowering the sphere into the vapor as done in these experiments. The disturbance of the vapor zone results in a transient period that is considerably longer, as discussed below, so that the thermal transient alone cannot be observed in the measured data.

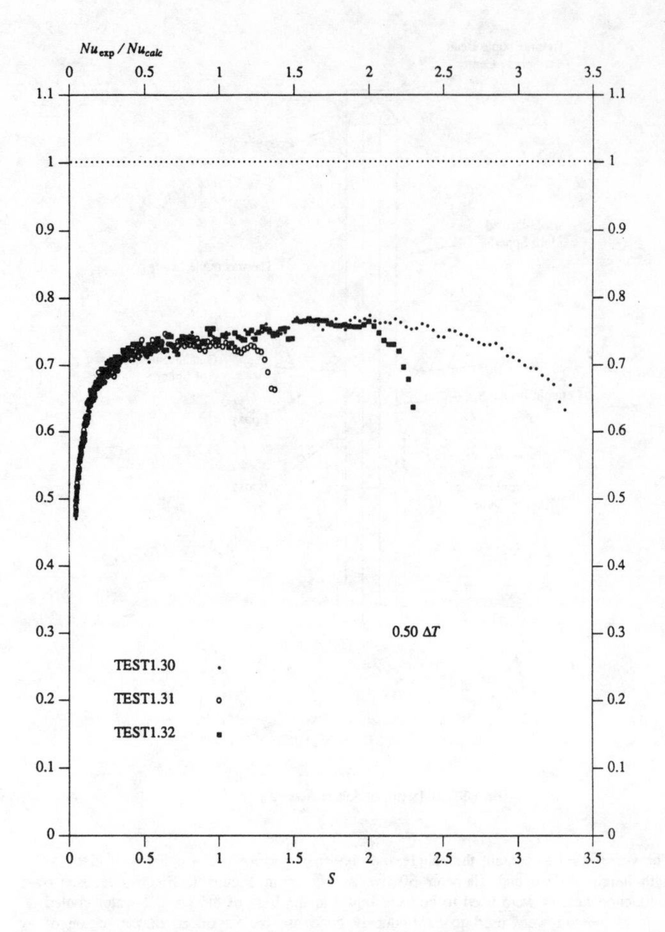

FIGURE 4. Comparison of Runs With Different Initial Sphere Temperatures.

When the interface between the vapor and air is disturbed, the vapor and air become inter-mixed creating a fog-like zone surrounding the entry region. This fog also follows the sphere into the vapor zone. Therefore, for the first few seconds of the experiment the condensation rate is expected to be lowered by the presence of air and possibly water vapor as non-condensable gas. A measure of the disturbance time is obtained by performing runs with elevated initial temperatures and observing the time required for the data to merge into that for the run with ambient initial temperature. Typical results are shown Figure 4. Runs TEST1.31 and TEST1.32 correspond to initial sphere temperatures of 100 ℃ and 150 ℃, respectively. For both the runs the insertion transients are clearly visible. They occur over a time period of approximately 4 seconds (8 data points), independent of initial temperature, and exhibit a very steep slope approaching the ambient initial temperature run. The decrease in the Nusselt number ratio at high S for the ambient initial temperature run occurs over a much longer period of time and is not solely an effect of initial transient. This is discussed in detail below.

5.2 Effect of Property Variations

The decline in Nusselt number ratio for Stefan number greater than about 2.3 can be attributed to the effect of viscosity variation in the condensate film. For example, for the test data shown in Figure 4, S = 2.3 corresponds to T_w = 96 ℃. The viscosity varies across the condensate film by a factor of about 5. For S = 3.35, corresponding to T_w = 30.8 ℃, the variation is a factor of about 34.

The standard way of correcting for property variations is to use a film temperature defined as

$$T_{ref} = T_w + C\Delta T \quad (12)$$

where the constant C takes on values between 0.23 to 0.33 (Minkowycz and Sparrow 1966, Poots and Miles 1967, Denny and Mills 1969). However, this correction is for water and the associated property variation is relatively small. The correction with C = 0.31 is shown in Figure 5. For larger property variation, the power law correction factor based on viscosity ratio is adapted to correct the calculated Nusselt number:

$$Nu_{vp} = Nu_{cp}\left(\frac{\mu_m}{\mu_w}\right)^n \quad (13)$$

where subscript m denotes the mean temperature (C = 0.5) across the boundary layer. This method has been successfully used for laminar forced flow in tubes. The exponent n takes on a value of either 0.11 (Yang, 1962) or 0.14 (Deissler, 1951).

With the power law correction using $n = 0.11$, the Nusselt number ratio approaches asymptotically a maximum value of about 0.87 at high S, as shown in Figure 5. Using the correction with $n = 0.14$, the ratio continues to rise with increasing slope. Note that the data corresponding to the initial disturbance period of 4 seconds has been included.

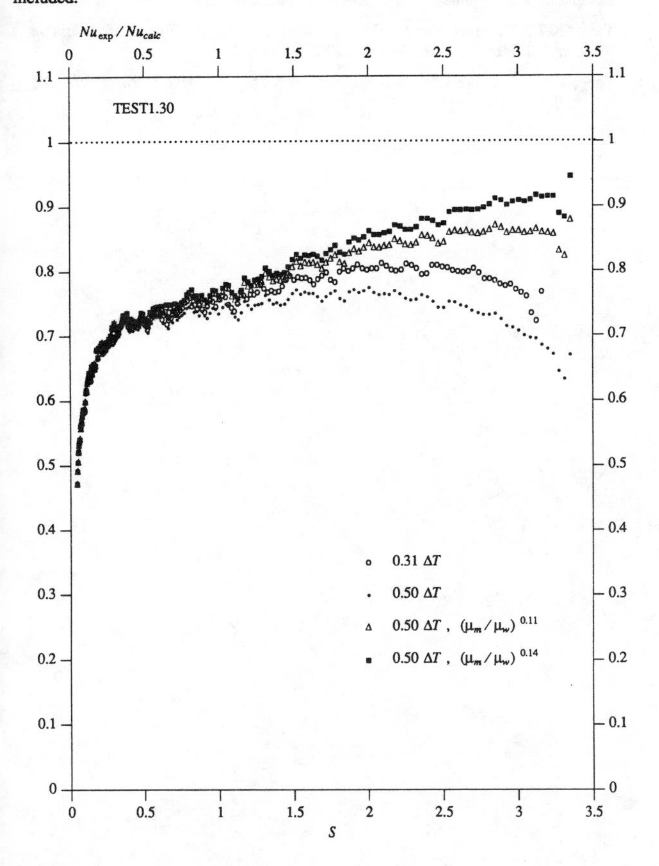

FIGURE 5. Effects of Reference Temperature and Viscosity Corrections

5.3 Effect of Noncondensable Components

The difference of about 10% between Nu_{exp} and Nu_{calc} using the power law correction (0.11 or 0.14) probably is due mainly to the presence of air which, as a noncondensible gas, impedes the motion of vapor to the condensing surface and hence reduces the rate of heat transfer. In addition, water vapor present at a concentration even as high as a 3% will have a condensing temperature less than 25 ℃ and hence also will be noncondensable over the full range of S.

The more or less linear decrease of the Nusselt number ratio with S in the range $0.3 < S < 3.3$ and the abrupt drop below $S \approx 0.3$ can be attributed to the fact that the working fluid FC-70 is not a single component fluid but a continuum of volatile components that condense over a range of temperatures. The results of distillation of a sample of FC-70 are plotted in Figure 6 in terms of temperature. The corresponding S is shown, assuming $T_{sat} = 215$ ℃. As the sphere temperature increases during an experimental run, an ever increasing fraction of the vapor becomes noncondensable, thus decreasing the rate of heat transfer. An inspection of Figure 6 shows that 1% of low temperature volatiles become noncondensable at approximately 203 ℃. For the sphere at this temperature, the corresponding value of S is 0.22. This is very close to the value of S below which the heat transfer rate drops precipitously. Relatively small amounts of noncondensables can have a significant effect on condensation rates (Rose, 1969 and Felicione and Seban, 1973).

Experimental results for the larger sphere, with diameter 50.8 *mm*, show the same trends as observed with the smaller sphere. Some representative heat transfer results are compared to those for the smaller sphere in Figure 7.

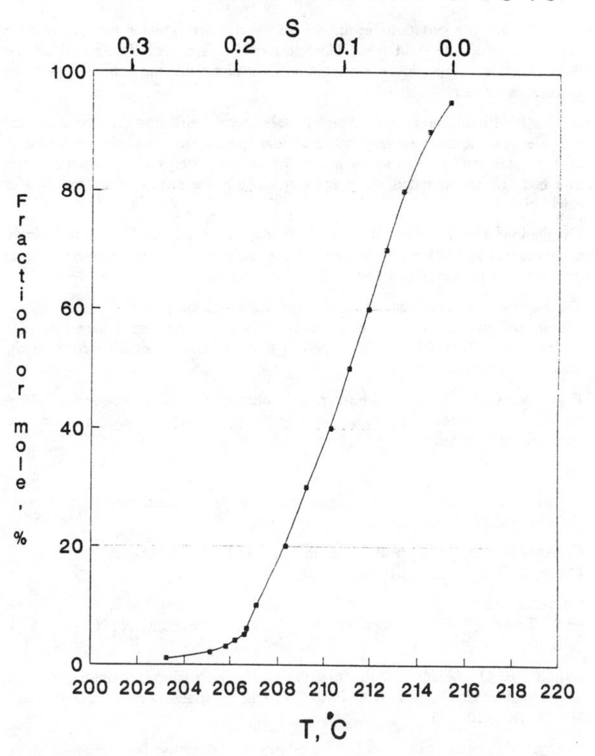

FIGURE 6. FC-70 Distillate Collected (% of Original Sample) as a Function of Temperature and Corresponding Stefan Number.

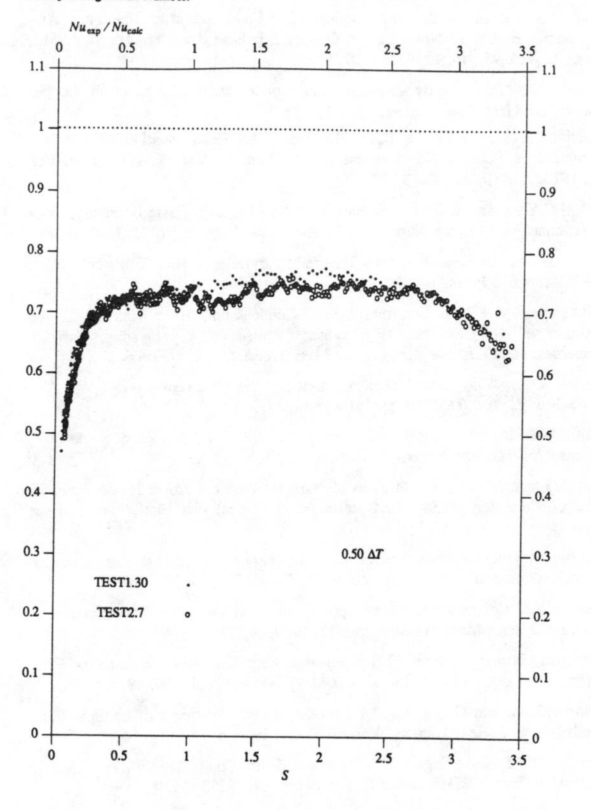

FIGURE 7. Comparison of Data for 1 inch (TEST1.30) and 2 inch (TEST2.7) Spheres

6. Concluding Remarks

1. Laminar film condensation results for Stefan numbers greater than unity have been presented. The fluid used, Flourinert FC-70 is one of the several similar fluids which are now routinely used in condensation soldering applications in the electronics industry.

2. For the fluid used, the ratio of dimensionless thermal diffusion time constant and sensible heat capacity is very small so that quasi-steady laminar condensation heat transfer results obtained by immersing copper spheres at room temperature in a body of hot saturated vapor are very close to the steady state heat transfer calculations.

3. Calculations indicate that for large Pr fluids, the simplified Nusselt-Rohsenow equations (1) and (2) recommended for use for $0 < S < 1$, can also be used to give accurate predictions of heat transfer rates for larger values of S.

4. The experimental heat transfer rates, when corrected for viscosity variations, are within 10% of the the theoretical values at large S. The deficit increases with decrease in S. This deficit is explained in terms of noncondensable components present in the vapor.

5. Experiments are planned in the future with ultra-pure single component distillate of FC-70 to further quantify the effect of the noncondensables on the experiments reported here.

References

1. Bromley, L.A., "Effect of Heat Capacity of Condensate", Ind. and Engr. Chem., Vol. 44, No. 12, 1952, p. 2966

2. Chen, M.M., "An Analytical Study of Laminar Film Condensation: Part I-Flat Plates", J. Heat Transfer 81, 1961, p. 48.

3. Chu, T.Y., Mollendorf, J.C., and Pfahl, R.C. Jr., "Soldering Using Condensation Heat Transfer," Proceedings of the Technical Program NEPCON, 1974, Anaheim.

4. Deissler, R. G., "Analytical Investigation of Fully Developed Laminar Flow in Tubes with Heat Transfer with Fluid Properties Variable Around the Radius", NACA TN 2410, 1951.

5. Denny, V.E. and Mills, A.F. "Non-Similar Solutions for Laminar Film Condensation on a Vertical Surface", Int. J. Heat Mass Transfer, Vol. 12, 1969, p. 965.

6. Dhir, V.K. and Lienhard, J.H., "Laminar Film Condensation on Plane and Axi-Symmetric Bodies in Non-Uniform Gravity", J. Heat Transfer, Trans. ASME, Series C, Vol. 95, No. 1, 1971, p. 97.

7. Dhir, V.K., "Quasi-steady Laminar Film Condensation of Steam on Copper Spheres", J. Heat Transfer, 1975, p. 347.

8. Felicione, F.S. and Seban, R.A., "Laminar Film Condensnation of a Vapor Containing a Soluble, Non-Condensing Gas," Int. J. Heat Mass Transfer, Vol. 16, 1973, p. 1601.

9. Koh, J.C.Y., Sparrow, E.M., and Hartnett, J.P., "The Two Phase Boundary Layer in Laminar Film Condensation", Int. J. Heat Mass Transfer, Vol. 2, 1961, p. 69.

10. Merte, H. Jr., "Condensation Heat Transfer", Advances in Heat Transfer, Vol. 9, 1973, Academic Press, p. 181.

11. Minkowycz, W.J. and Sparrow, E.M., "Condensation Heat Transfer in the Presence of Non-Condensables, Interfacial Resistance, Superheating, Variable Properties, and Diffusion", Int. J. Heat Mass Transfer, Vol. 9, 1966, p. 1125.

12. Nusselt, W., "Die Operflachenkondensation des Wasserdampfes," Z. Ver Deutsch, Ing., Vol. 60, 1916, pp 541 and 569.

13. Pfahl, R.C. Jr., Mollendorf, J.C., and Chu, T.Y., "Condensation Soldering," Welding Journal, Vol. 54, No. 1, 1975, p. 22

14. Poots, G. and Miles, R.G., "Effects of Variable Physical Properties on Laminar Film Condensation of Saturated Steam on a Vertical Flat Plate," Int. J. Heat Mass Transfer, Vol. 10, 1967, p. 1677.

15. Rohnsenow, W.M., "Heat Transfer and Temperature Distribution in Laminar Film Condensation," Trans Am. Soc. Mech. Engrs. 78, 1956, p. 1645.

16. Rose, J.W., "Condensation of a Vapor in the Presence of a Non-Condensing Gas", Int. J. Heat Mass Transfer, Vol. 12, 1969, p. 233.

17. Sadasivan, P. and Lienhard, J.H., "Sensible Heat Correction in Laminar Film Boiling and Condensation", Int. J. Heat Mass Transfer, Vol. 109, 1987, p. 545.

18. Sparrow, E.M. and Gregg, J.L. "A Boundary-Layer Treatment of Laminar-Film Condensation, J. Heat Transfer, Vol. 81, No. 1, 1959, p. 13.

19. Sparrow, E.M. and Siegel, R., "Transient Film Condensation," J. Applied Mechanics, Trans. ASME Series E, Vol. 81, No. 1, 1959, p. 120.

20. Wenger, G.M. and Mahajan, R.L., "Condensation Soldering Technology - Part I: Condensation Soldering Fluids and Heat Transfer", Insulation/Circuits, Sept. 1979, p. 131.

21. Wenger, G.M. and Mahajan, R.L., "Condensation Soldering Technology-Part II: Equipment and Production", Insultation/Circuits, Oct. 1979, p. 133.

22. Wenger, G.M. and Mahajan, R.L., "Condensation Soldering Technology-Part III: Installation and Application", Insulation Circuits, Nov. 1979, p. 13.

23. Yang, K.T., "Laminar Forced Convection of Liquids in Tubes with Variable Viscosity", J. Heat Transfer, Vol. 84, 1962, p. 353.

24. Yang, J.W., "Laminar Film Condensation on a Sphere", Heat Transfer, Vol. 95, No. 2, 1973, p.174